International Submarine Cables and Biodiversity of Areas Beyond National Jurisdiction

International Submarine Cables and Biodiversity of Areas Beyond National Jurisdiction

The Cloud Beneath the Sea

By

Douglas R. Burnett and Lionel Carter

BRILL

LEIDEN | BOSTON

Originally published as Volume 1(2) 2017, in *Law of the Sea*, DOI 10.1163/24519359-12340002.

Library of Congress Control Number: 2017943572

Typeface for the Latin, Greek, and Cyrillic scripts: "Brill". See and download: brill.com/brill-typeface.

ISBN 978-90-04-35159-2 (paperback)
ISBN 978-90-04-35160-8 (e-book)

This book is printed on acid-free paper and produced in a sustainable manner.

Contents

Author Biographies

Douglas R. Burnett

International Law Advisor for the International Cable Protection Committee (since 1999) and Maritime Partner in the New York office of Squire Patton Boggs (US) LLP, an international law firm with 44 offices in 21 countries. His practice focuses on international law, submarine cables, maritime and shipping law, involving litigation and arbitration. Douglas is a graduate of the US Naval Academy and University of Denver Law School and is a retired captain in the US Navy. He has argued before the U.S. Supreme Court and testified as an industry expert on submarine cables before the 2007 Senate Foreign Relations Committee for the United Nations Convention on the Law of the Sea (UNCLOS) hearings. Douglas has worked on submarine cable cases for over 35 years. He is co-editor and co-author of the treatise *Submarine Cables:The Handbook of Law and Policy* (Brill, 2014) and a contributing author of the *Routledge Handbook on Ocean Resources and Management* (2015), and over 36 articles on submarine cables, maritime and admiralty topics, and law of the sea. He has frequently instructed at the Rhodes Academy of Oceans Law and Policy in Rhodes, Greece.

Lionel Carter

Marine Environmental Advisor for the International Cable Protection Committee (since 2003) and Professor of Marine Geology, Antarctic Research Centre, Victoria University, Wellington, New Zealand. Lionel trained in geology and oceanography at the universities of Auckland, New Zealand and British Columbia, Canada, and has undertaken research in the North Pacific, North Atlantic and Southern oceans, as well as off New Zealand. This research led to publication of over 140 peer-reviewed papers. He helped set up major international projects in New Zealand, including Ocean Drilling Program Leg 181 and the MARGINS "Source to Sink" initiative, which was designed to determine the processes that shape Earth's surface from the mountains to the abyssal ocean. Lionel's present research centres on deciphering marine geological records to identify changes in ocean/climate. This work contributes to the development of observational and numerical models that aid prediction of potential environmental responses to the on-going warming climate. His expertise is also applied to ocean engineering projects, in particular submarine telecommunication and power cables. He was elected as a Fellow of the Royal Society of New Zealand in 2003.

Disclaimer: The opinions of the authors are theirs alone and do not reflect the views of the International Cable Protection Committee or its members, Victoria University of Wellington, or Squire Patton Boggs (US) LLP.

International Submarine Cables and Biodiversity of Areas Beyond National Jurisdiction

The Cloud Beneath the Sea

Douglas R. Burnett
Squire Patton Boggs (US) LLP
New York, NY, USA
douglas.burnett@squirepb.com

and

Lionel Carter
Antarctic Research Centre, Victoria University
Wellington, New Zealand
lionel.carter@vuw.ac.nz

Abstract

If one uses Facebook, Facetime, Skype, Netflix, or any application of the Internet internationally, a submarine cable is involved. Fibre-optic cables bind the world together from governments, banks, shipping, airlines and other major logistic industries to homes and personal electronic devices. Server farms maintained by major telecom and content companies allow vast amounts of data to be stored and retrieved from the *cloud*. Not often appreciated is the fact that these server locations worldwide are connected by submarine fibre-optic cables. In this sense, the *cloud* is beneath the sea. While submarine communication cables have been in steady use since 1850, their preeminent place in the modern world has never been as dominant and personal as now. Since 1884, this critical international infrastructure has rested upon international treaties, now reflected in universally accepted provisions of the 1982 United Nations Convention on the Law of the Sea (UNCLOS) that provides for freedoms to lay and maintain international submarine cables. Recently, calls have mounted in the context of marine biodiversity beyond national jurisdiction (BBNJ) for centralized control of submarine cables and for express or *de facto* diminishment of the freedoms related to them that have served the world's peoples for so long. This monograph examines the time proven importance of the existing international treaties, the largely peer review science on the environmental interaction of submarine cables with high seas

 | DOI 10.1163/9789004351608_002

environments, and the current submarine cable issues in the context of the BBNJ debates.

Keywords

submarine cables – UNCLOS – marine environment – precautionary approach – high seas – BBNJ – PrepCom

I Introduction

The Oceans and the Law of the Sea Report of the Secretary-General,[1] in its paragraphs 53 and 55 succinctly sums up the conventional wisdom about international submarine cables and sustainable development:

> Submarine cables are critical communications infrastructure, being used for more than 98 per cent of international internet, data and telephone traffic, with only a few States without fibre connectivity, and many of these having cable projects currently under way.[85][2] Submarine cables are recognized as vitally important to the global economy[86][3] and hence to economic growth. By underpinning international communications, their role in providing access to data and information for all peoples is evident. /…/
>
> Functioning as the backbone of the international telecommunications system, submarine cables are a fundamental component of the critical global infrastructure and play a direct role in sustainable industrialization; indirectly they contribute to all other areas recognized as important for sustainable development.

Amazingly, when people think about international communications, they often mistakenly regard satellites as the primary communications medium. Many

1 UN doc. A/70/74 (30 March 2015).

2 Ibid., footnote 85 in the report, which reads: 'D. Burnett, D. Freestone and T. Davenport, "Submarine cables in the Sargasso Sea: legal and environmental issues in areas beyond national jurisdiction", report of a workshop held in Washington, D.C., on 23 October 2014 (2015)'; the report is available from the Sargasso Sea Commission (http://www.sargassoseacommission.org/) or the International Cable Protection Committee (www.iscpc.org/).

3 Ibid., footnote 86 in the report, which reads: 'See General Assembly resolution 69/245'.

express surprise to learn that over 98% of international communications are carried by a relatively small number of fibre-optic submarine cables with diameters akin to a domestic garden hose even though such systems have been operational for some 30 years (i.e., since the onset of the fibre-optical era, 1987 to the present).

The confusion is understandable. The idea that a person's cell phone link is sent to a nearby cell tower, but that the overseas messages themselves are then broken into bits of data, which then ply the ocean depths at the speed of light via unseen cables, is hard to imagine. The tremendous volume of data carried at low cost by modern fibre-optic submarine cables dwarfs the speed and limited capacity of higher cost satellites. For example, the capacity of a single transatlantic cable has increased by a factor of 100,000 in 25 years.[4] Additionally, the technical transmission delays, modest capacity and other quality limitations inherent in satellites make them comparatively marginal for continuous transmission of high-speed voice, video and data traffic.

The collective impact of the laypersons' mistaken beliefs and knowledge gap is negatively compounded by the fact that many in government share their misconceptions, even as they fashion ocean policies and regulations that overlook submarine cable history, marine engineering, seamanship, environmental aspects and international law. Often, these flawed regulatory efforts undercut the viability of the successful submarine cable network as the critical international infrastructure upon which the internet and global economy are based. Like cables, the time tested and very successful international legal regime that supports international cables is often misunderstood or overlooked.

Even more unknown or appreciated is the substantial body of scientific research and papers that document the inter-relationship of submarine cables and the marine environment.

Also little recognized is the critical role that international law has played and continues to play in the innovation fostered by the freedom to lay and maintain international cables. Given that the demand for internet capacity is increasing at a rate of about 40% per year,[5] understanding this symbiotic relationship between the law and cables is both vital and urgent.

Submarine cable systems do not exist in a vacuum. They are the careful product of collaboration by an international cable community that includes

4 J. Chesnoy, 'Back Reflection' (2016) 88 *Subtelcom Forum*, at p. 96; 'Geomesh: Global Submarine Networking for a Web-scale World' (2016) *Ciena White Paper*, at p. 1, predicting total international capacity demand in 2020 will be around four times the 2015 level.

5 'Peak Mobile Bandwidth Per User To Increase Fivefold By 2018', *Ciena Study*, 2 December 2015, at: http://media.ciena.com/documents/2015_02_12_Ciena_ACG_Research_Mobile_FINAL.pdf (last accessed on 28 November 2016).

cable owners, cable ship operators, marine route surveyors, scientific institutions and interested governments (i.e., collectively, the "submarine cable community", as referred to hereinafter).

The purpose of this monograph is to provide useful up-to-date background information for scholars, diplomats, ocean policy makers, scientists, and others about submarine cables, their role in sustainable development, their presence in the marine environment, and the relationships with international law. The monograph is organized in four sections. The introduction section provides background on submarine cables, economic and social development and science. The second section provides an overview of international law applicable to submarine cables. The third section reviews submarine cables in the marine environment as reported in peer review articles, international workshops, and general literature. The last section addresses the adequacy of existing international law on submarine cables in the context of new ideas being debated for BBNJ governance.

A *Submarine Cables and Economic and Social Development*

Each day the Society for Worldwide Interbank Financial Telecommunications (SWIFT) transmits 15 million messages over cables to over 8,300 banking organizations, securities institutions and corporate customers in 208 countries. The Continuous Linked Settlement (CLS) Bank located in the United Kingdom is just one of the critical market infrastructures that rely on SWIFT as it provides global settlement of 17 currencies with an average daily US dollar equivalent of approximately USD 3.9 trillion. The US Clearing House Interbank Payment System (CHIPS) is another structure that processes over USD 1 trillion a day to over 22 countries for investment companies, securities and commodities exchange organizations, banks and other financial institutions.[6]

If the approximately 40 or so garden-hose size cables connecting the United States to the rest of the world were cut, even using every single satellite in the sky, it is estimated that only 7% of the total United States traffic volume could be carried by satellite.[7] Referring to the submarine cable networks, the Staff Director for Management of the Federal Reserve observed "when the communication networks go down, the financial sector does not grind to a halt,

6 S. Malphrus, 'Undersea Cables and International Telecommunications Resiliency', 34th Annual Law of the Sea Conference, Center for Ocean Law and Policy, University of Virginia, 20 May 2010.

7 The testimony of D. Burnett before the Senate Foreign Relations Committee on the United Nations Law of the Sea Convention, 4 October 2007.

it snaps to a halt."[8] The same can be said for most industries enmeshed in the global economy through the internet including shipping companies, airlines, banks, supply chain, manufacturing businesses, and entertainment. This underscores the fact that if something were to happen to cause the loss of submarine cables, there is no "Plan B" available to replace the international capacity they carry.

Other countries are no different in their reliance. Australia and Singapore, for example, each rely on several cables landing in each nation for over 99% of their international communications. Japan does the same with about 20 international cable systems. And the list goes on. With the laying of submarine cables along the east coast of Africa in 2009–2010, the last major group of nations gained access to the world's submarine cable network. As of mid-2012, only 22 nations and territories remained isolated from fibre-optic connectivity and many of these have connecting cable projects underway.[9] A major challenge now being met by the submarine cable community is providing connections to small island economies and isolated coastal communities of the high Arctic,[10] together with provision of redundant cable connections to keep these economies connected in the event of a cable fault.

The world's dependence on reliable low cost and secure submarine cables continues to grow: "Every *second* they can carry 31 terabits across the Pacific and 55 terabits across the Atlantic."[11] A look at the websites of major companies like Google, Microsoft, Facebook and Amazon shows the diverse locations

8 S. Malphrus, Board of Governors of the Federal Reserve System, First Worldwide Cyber Security Summit, East-West Institute, Dallas, Texas, 3–5 May 2010.

9 Submarine Telecoms Forum, Inc., *Telecoms Industry Report 2012*, at pp. 14–15. Inhabited sovereign States and territories without fibre optic connectivity include: Somalia, Saint Helena, Ascension, and Tristan da Cunha (British Overseas Territory); Christmas Island (Australian External Territory), Montserrat (British Overseas Territory); Saint Pierre and Miquelon (French collectivité d'outre-mer); Easter Island (Chilean Special Territory), Falkland (Malvinas) Islands (British Overseas Territory), Cook Islands (Self-Governing State in Free Association with New Zealand), Kiribati, Nauru, Niue (Self-Governing State in Free Association with New Zealand, Norfolk Island (Australian External Territory), Palau, Pitcairn Islands (British Overseas Territory), Solomon Islands, Tokelau (New Zealand Dependent Territory), Tonga, Vanuatu, Wallis and Futuna (French collectivité d'outre-mer); Timor Leste.

10 Installation of the first Arctic cable system [Quintillion, 2016] began in Alaska during the summer of 2016 and is scheduled to complete installation of phase 1 in 2017. Phases 2 and 3 will provide for this international cable system to be connected from the United Kingdom to Japan with landings in Alaska and Canada via the Northwest Passage. http://qexpressnet.com/system/ (last accessed on 28 November 2016).

11 'The See-through Sea', *The Economist*, 16 July 2016, at p. 16.

of the legions of computer servers in each company's data centres which are distributed worldwide and on every inhabited continent except Africa. These *cloud* data centres are seamlessly connected by international submarine fibre-optic cables. So critical to their business success are submarine cables that these companies have all become cable owners in addition to leasing capacity on other cable systems. It is not an exaggeration to say the *cloud* would not exist but for cables under the sea.

The future contributions of submarine cables to sustainability continue to expand. Technology advances in 3D printing will allow items to be locally manufactured, reducing the need for ocean or air transportation and other high carbon footprint industries. For example, an electric lightweight 80 km/h motorcycle entered print production in 2016.[12] International fibre-optic cable connections allow the software, blue prints, and payments to flow. Similarly, long distance learning allows dispersed populations in archipelagic States to have common access to the best teachers and resources. Medical data and expertise shared by cables allow for improved health care and bring high-level care to more and more people.

By 2020, one expert estimates that there will be 4 billion people connected to the internet, USD 4 trillion of revenue opportunity from these connections, using over 25 million apps, with over 25 billion embedded intelligent systems, and 50 trillion gigabits of data.[13] These connections will exist almost exclusively on international submarine cables, the backbone of the internet. Applications such as Skype, Facetime, Netflix, Twitter, Facebook, and YouTube remind us all in a personal way that our lives are directly affected by submarine cables.

B *Submarine Cables and Science*

Besides the critical economic and social sustainability roles played by the submarine cables as international arteries connecting the world, submarine cables are also vital for marine science and the quest to learn more about the oceans and climate. In a 2009 survey, the International Cable Protection Committee (ICPC)[14] identified 193 ocean observation sites and areas worldwide, including

12 'Alloy Angels', *The Economist*, 28 May 2016, at p. 72.

13 Source: Mario Morales, Vice-President, IDC Research Inc highlighted in Telegeography Webinar 'New Cable Builds Return to the Trans-Atlantic: What Are the Impacts?', 6 October 2015.

14 The ICPC is an organization of 166 members from 60 countries (status as of 6 June 2017) that own and operate over 98% of the world's international submarine cable telecommunication systems, as well as several high voltage direct current (HVDC systems) and the majority of the cable ships in the world that lay and repair these systems. Membership

at least 34 that planned or were using submarine cables for data transmission and power transfer in the world's oceans.[15] The 800 km cable-based Neptune system, with multiple scientific nodes (special seabed housings capable of supporting monitoring equipment and experiments) off Vancouver Island, British Columbia is a standout operational example. Another is the Ocean Observatory Initiative (OOI) array off Washington-Oregon, which like its Canadian counterpart is based on a 900 km cable that supports a suite of nodes.[16] Japan has pioneered the use of submarine cable systems to monitor seismic activity and detect tsunamis.[17]

II International Law and Submarine Cables

Historically, several treaties have addressed submarine cables. In fact, the first ever "law of the sea" treaty—the 1884 International Convention for the Protection of Submarine Cables—dealt solely with submarine telegraph cables.

A *The 1884 International Convention for the Protection of Submarine Cables*

Diplomatic conferences were held to create an international treaty to protect and foster the growth of the then revolutionary new technology of submarine telegraph cables—the "internet" of that day. The fascinating account of these conferences is captured in the notes of Professor Renault, the French scholar

---

is available to Submarine Cable Owners, Submarine Cable Maintenance Authorities, Submarine Cable System Manufacturers, Cable Ship Operators, Submarine Cable Route Survey Companies and Governments. For a list of ICPC membership see: https://www.iscpc.org/about-the-icpc/member-list/. Governments have been members of ICPC since 2010; there are presently five such members: Australia, Malta, New Zealand, Singapore, and the United Kingdom. For additional information, see the ICPC website at www.iscpc.org (last accessed on 1 March 2017).

15 ICPC Ocean Observation Sites and areas-2009; see at www.iscpc.org. The survey results were compiled by the author (Dr. Lionel Carter).

16 Ocean Networks Canada, 2016. *Discover the Ocean, Understand the Planet: Observatories* at: http://www.oceannetworks.ca/; and OOI, 2017. *The OOI Cabled Network*, at: http://www.interactiveoceans.washington.edu/story/ (last accessed on 1 March 2017).

17 C. Manoj, A. Kuvshinov, S. Neetu and T. Harinarayana, 'Can undersea voltage measurements detect tsunamis?' (2006) 58 *Earth Planets Space* 1–11; R. Monastersky, 'The next wave' (2012) 483 *Nature* 144–146.

considered the *de facto* rapporteur of the conferences.[18] Because the technology was pioneering, besides diplomats, there was effective participation at the conference by engineers, fishermen, and naval officers from various countries. The remarkable result is the 1884 International Convention for the Protection of Submarine Cables ("Cable Convention").[19] The Cable Convention is the bedrock for the provisions of modern international law on submarine cables found in UNCLOS.[20]

The Cable Convention is the first widely accepted treaty to establish the freedom to lay and maintain international cables outside of territorial seas.[21] The Cable Convention requires all parties to enact domestic laws making the wilful or culpably negligent injury to a submarine cable a crime and to provide for enforcement in national courts.[22]

The Cable Convention is best known for the innovative compromises among the parties that resulted in provisions carried over into UNCLOS based on their proven practical utility over time since 1884.

The first compromise is found in Article 2 that balances the need to protect with criminal and civil sanctions submarine cables from willful injury or injury caused by culpable negligence against the actions of a master taken to save his vessel and crew.

The second compromise, found in Article 4, is the resolution of cable crossings. If two cables cross and both companies enjoy the freedom to lay their cable, who bears the loss if a cable is damaged in crossing? The compromise is that the priority lies with the first laid cable. While every cable can cross another cable, if in the course of the crossing the first laid cable is damaged, the crossing cable must indemnify the first laid cable for the cost of repairs.[23]

---

18 L. Renault, 'The Protection of Submarine Telegraphs and the Paris Conference (October–November 1882)', in *International Law Review*, ed. Brussels and Leipzig, (Flander: Merzbach and Falk, 1884), (hereinafter: Renault 1884). Note: the International Cable Protection Committee (ICPC) in its member legal data base has English translations of many contemporaneous records of the Paris Conference. Access to scholars is freely available upon request to the Managing Director.

19 14 March 1884, T.S. 380. The provisions of the Cable Convention are generally accepted as customary international law. *Restatement (Third) of the Foreign Relations Law of the United States* § 521, comment f (1986). There are 41 State parties to the Cable Convention.

20 United Nations Convention on the Law of the Sea (Montego Bay, 10 December 1982; in force 16 November 1994), 1833 *UNTS* 396.

21 Article 2, Cable Convention.

22 Articles 8–10, Cable Convention.

23 Article 4, Cable Convention, and Renault 1884 (n. 18), at p. 11.

In modern practice, companies planning a new cable system go to great lengths to identify any cable or pipeline that must be crossed. Advance liaison is carried out to plan for a safe crossing. In some cases there is a voluntary formal crossing agreement. In others there is not because there is no requirement for such an agreement under international law. In any event, crossing arrangements typically receive careful engineering and planning scrutiny and liaison with both systems involved in a crossing so that damages to cables and pipelines are rare.

The third compromise, found in Article 7, is the situation of a vessel or master, which for no fault of its own finds it has fouled a cable with its fishing gear or anchor. In this circumstance, the Cable Convention requires the vessel to sacrifice its gear or anchor to avoid the greater harm of disrupting international communications.[24]

In return the Cable Convention requires that the cable owner indemnify the vessel for the replacement cost of the sacrificed gear.[25] The Cable Convention then details the procedure for a vessel to claim indemnity by filing witness statements and cost supports with the cable owner, or if not known, with the captain of the port or the coast guard within 24 hours of arrival in port. This practice is widely followed by cable owners who generally maintain 24-hour telephone hotlines to receive reports and provide information to masters of vessels who report they may be fouled on a cable. Once a claim is filed, the cable owner investigates and in most cases the sacrificed gear is recovered and returned or indemnity compensation is paid. As required by international law, the indemnity is limited to the costs of the actual sacrificed gear or anchor and does not include damages for lost profits or catch.[26]

Each of the three compromises described has been included in modern form in the 1958 Geneva Convention on the High Seas[27] and UNCLOS with cable provisions that track those of the 1884 Cable Convention. These relationships are displayed in Table 1:

24 Article 7, Cable Convention, and Renault 1884 (n. 18), at p. 13.

25 Ibid.

26 Article 7, Cable Convention, and Renault 1884 (n. 18), at p. 13; and *Agincourt Steamship Company Ltd. v Eastern Extension, Australia and China Telegraph Company Ltd.* [1907] 2 K.B. 305 (United Kingdom).

27 29 April 1958, 450 *UNTS* 11 (entered into force 30 September 1962); hereinafter 'High Seas Convention'.

TABLE 1 *Comparative overview of the 1982 UNCLOS, the 1958 Geneva High Seas Convention and the 1884 Cable Convention Articles*

UNCLOS Article	Geneva Convention Article	1884 Convention Article
113	27	2
114	28	4
115	29	7

There are, however, articles in the Cable Convention dealing with navigational safety and boarding that are not found in UNCLOS.

Article 5 requires that vessels maintain a distance of one nautical mile (1.85 km) from a cable ship laying cable, which displays the appropriate day shapes or lights at night denoting its restricted manoeuvrability status to other vessels in the area.[28] Article 6 requires vessels to maintain a safety distance of one fourth of a nautical mile (0.46 km) from a cable repair buoy.[29]

To some extent the restricted navigation status of a cable ship engaged in laying or repairing a cable is addressed in the 1972 International Regulations for Preventing Collisions at Sea (COLREGS).[30] But modern experience highlights sharply that the COLREGS provisions need to be re-examined to address the increasing interference of fishing vessels with cable ships and cable repair buoys, and to reinforce in COLREGS a safety distance of one nautical mile from a working cable ship and one quarter of a nautical mile from a cable repair buoy.[31] Serious maritime safety issues involving fishing vessels that often ignore safety distances and impede laying and repair of cables are common in Asian waters. Attempts in 2014–2015 by the United States and the Marshall Islands to galvanize the International Maritime Organization (IMO) to issue IMO guidelines or issue a unified interpretation of COLREGS to provide for inclusion of the 1884 safety distances from cable ships and repair buoys in the

28 Article 5, Cable Convention, and Renault 1884 (n. 18), at pp. 11–12.

29 Article 6, Cable Convention, and Renault 1884 (n. 18), at p. 12.

30 Rules 3(g)(i), 18, and 27, COLREGS.

31 D. Burnett, 'The 1884 International Convention for Protection of Submarine Cables Provisions Not in UNCLOS Deserve Attention Now', Workshop on the Protection of Submarine Cables, Singapore, 14–15 April 2011, sponsored by the Centre for International Law, National University of Singapore and the International Cable Protection Committee, http://cil.nus.edu.sg/powerpoint-presentations-and-papers-presented-at-the-2011-workshop/.

Rule 18 [Responsibilities between vessels] obligation to "keep out of the way", were not successful.[32] While the overwhelming majority of the State representatives were in favour of such action, a few States led by Panama, in the consensus environment fostered by IMO, blocked this needed action to safeguard critical international infrastructure.[33]

Another provision of the 1884 Cable Convention not found in UNCLOS is Article 10, which allows an officer from a naval vessel from any of the parties to board a ship on the high seas suspected of damaging an international cable.[34] This provision was used in 1959 by a USS Navy destroyer, which boarded a Soviet trawler suspected of cutting five transatlantic cables over a two-day period.[35]

Under UNCLOS, boarding of vessels by warships or coast guard vessels outside territorial seas is restricted. A vessel reasonably suspected of engaging in piracy or refusing to show its flag can be boarded.[36] Similarly, vessels can be boarded with flag state consent. An argument can be made that Cable Convention parties have already consented for visits by warships of other parties. But these nations constitute a small number of the world's merchant and fishing fleets. UNCLOS does not provide for boarding save by flag State consent of vessels suspected of hostile acts against submarine cables or cable ships.

32 IMO Sub-Committee on Navigation, Communications and Search and Rescue document NCSR 3/1/2, 14 January 2016, at: https://edocs.imo.org/Final Documents/English/NCSR 3-1-2.

33 NCSR 3/WP.4 Report of the Navigation Working Group (3 March 2016) [5. Interpretation of COLREG Rule 18: Protection of Cable Ships], at pp. 7–9.

34 Article 10, Cable Convention, and Renault 1884 (n. 18), at p. 18.

35 The Novorossiisk, *Dept. of State Bull.*, Vol. XL, no. 1034, at 555 (20 April 1959). (www.iscpc.org/members/Legal/Precedence Cases/Precedence_Case_File_12.pdf). Press release: The Embassy of the United States of America refers to the Ministry's note No. 17/OSA, dated 4 March 1959 concerning recent breaks in certain transatlantic submarine telecommunication cables and the consequent visit to the Soviet trawler *Novorossiisk* by a boarding party from the *USS Roy o. Hale*, which was the subject of the Embassy's aide memoire of 28 February 1959.

36 Articles 101(a)(ii) and 110, UNCLOS; M. Green and D. Burnett, 'Security of International Cable Infrastructure: Time to Rethink?', *Challenges in Maritime Security*, Center for Oceans Law and Policy (2008).

B The 1958 Geneva Conventions on the High Seas and on the Continental Shelf

The 1958 Geneva Conventions contain provisions on both the protection of submarine cables on the high seas and the freedom to lay submarine cables in the high seas and on the continental shelf.[37]

Ultimately, only three articles on the protection of submarine cables in the 1884 Convention were incorporated into the High Seas Convention.[38] In order to satisfy parties to the 1884 Cable Convention that nothing in the High Seas Convention would undercut the earlier convention, Article 30 was agreed to remove any doubt on this point.[39]

Building on the foundation of the 1884 Cable Convention, the High Seas Convention expanded cable protection in several ways. Under the earlier convention, no action could be taken against wrongdoers until after the cable was injured. Under the High Seas Convention, a State can also take action to protect cables from actions liable to interrupt cables.[40] Moreover, the 1884 object of telegraph cables was expanded to include telephonic as well as high-voltage power cables.[41]

Coastal States are prevented from laying cables or pipelines in such a manner as to prejudice the repair of existing cables or pipelines.[42] The 1958 Geneva Convention on the Continental Shelf[43] contains a similar provision providing for the freedom to lay and maintain cables upon the continental shelf taking into account the coastal States' reasonable rights of exploration and exploitation of natural resources.[44]

37 450 *UNTS* 11 (High Seas Convention), Articles 27–29; and 499 *UNTS* 311 (Continental Shelf Convention), Article 4.

38 Articles 2, 4 and 5 of the 1884 Convention were incorporated in Articles 27, 28 and 29 of the 1958 High Seas Convention.

39 *Conventions on the Law of the Sea: Hearing before the Committee on Foreign Relations, United States Senate, 86th Cong., 2nd Sess., 20 January 1960* (US Government Printing Office: Washington DC, 1960), at pp. 92, 106.

40 Article 27, High Seas Convention.

41 Article 27, High Seas Convention.

42 Article 26, High Seas Convention.

43 29 April 1958, 499 *UNTS* 311 (entered into force 10 June 1964), hereinafter 'Continental Shelf Convention'.

44 Article 4, Continental Shelf Convention.

C *The 1982 United Nations Convention on the Law of the Sea*

The Preamble to UNCLOS emphatically recognizes "the desirability of establishing through this Convention ... a legal order for the seas and oceans which will facilitate international communication." Perhaps no part of UNCLOS better carries out this object than its articles dealing with submarine cables. Articles 21(1)(c), 51(2), 58, 78, 79, 87(1)(c), 112, 113, 114, 115 and 297(1)(a) fulfil the object and purpose of facilitating international communication.

Under UNCLOS the many uses of submarine cables besides international communication are recognized. These include cables used to exploit conventional natural resources (cabled oil and gas production platforms), alternative energy (off-shore wind farms, wave and tidal current generators), marine scientific research (cabled ocean observatories and ocean monitoring systems), international High Voltage Direct Current ("HVDC") power cables between States and cables used for military purposes. Except as qualified in UNCLOS, all of these cable uses in an international context enjoy the same rights and obligations.

The provisions on the protection, laying and repair of submarine cables in the 1958 Geneva Conventions are reproduced more or less *ad verbatim* in UNCLOS, which presently has 168 parties[45] and prevails, as between States Parties over the 1958 Geneva Conventions.[46] The UNCLOS provisions on submarine cables represent customary international law and are hence binding on non-parties.[47]

In reviewing UNCLOS the standard is that it is "interpreted in good faith in accordance with the ordinary meaning to be given to terms of the treaty in their context and in the light of its object and purpose."[48] The "context" refers

---

45 Status of UNCLOS, UN Treaty Collection, available at the UN Treaty Collection website, https://treaties.un.org/Pages/ViewDetailsIII.aspx?src=TREATY&mtdsg_no=XXI-6&chapter=21&Temp=mtdsg3&clang=_en#4.

46 Article 311(1), UNCLOS.

47 Most of the UNCLOS provisions on submarine cables are based on the provisions found in the 1958 High Seas Convention which purported to codify existing customary international law; see R.R. Churchill and A.V. Lowe, *The Law of the Sea*, 3rd edn (Manchester University Press, 1999), p. 203. With regards to the provisions in UNCLOS not based on the 1958 High Seas Convention, UNCLOS in general is almost universally accepted and the majority of its provisions would also be best evidence of customary international law; see Churchill and Lowe, ibid. The United States has taken the position that UNCLOS reflects customary international law to which the United States adheres; President Ronald Reagan Statement on United States Ocean Policy, 19 *Weekly Comp. Pres*, doc. 383 (10 March 1983).

48 Article 31, Vienna Convention on the Law of Treaties (23 May 1969).

only to the actual terms of the treaty, including its preamble. UNCLOS is a carefully balanced package of compromises. What is not stated in UNCLOS can be as important as what is stated. The aspirational tendency of some to imply through expansive interpretive logic additional conditions, meanings, or terms to supplement the agreed text should be avoided.[49] Many ideas were debated during the years of negotiating UNCLOS, but in the long diplomatic process were dropped or modified in favour of the stated compromises in the text.

UNCLOS establishes legal regimes for the territorial seas, the exclusive economic zone (EEZ), the continental shelf, and the high seas.

1 Territorial Sea and Archipelagic Waters

Territorial seas may not exceed 12 nautical miles from the coastal baseline. Article 21(1)(c) [Laws and regulations of the coastal State relating to innocent passage] allows the coastal State to adopt laws and measures for "the protection of cables and pipelines" which may limit innocent passage of vessels within territorial seas.[50] This is consistent with the sovereignty that coastal States enjoy in their territorial seas that allows them to set the conditions of cables within the territorial sea, including denial of landings and transit for international cables.

Given the importance of submarine cables, most nations have detailed regulations for any cable system that seeks to land in a State or transit its territorial sea. Article 21(1)(c) measures employed include the establishment of corridors around submarine cables where dredging, anchoring and other seabed activities, which can damage cables, are restricted. Australia, New Zealand, Denmark, Uruguay, and Colombia have modern and effective domestic laws in national waters that establish protection zones around international cables that land in those countries.

UNCLOS recognizes archipelagic waters and accords archipelagic States rights and obligations similar to those a coastal State enjoys in its territorial seas. Article 51(2) [Existing agreements, traditional fishing rights and existing submarine cables] requires an archipelagic State to "respect existing submarine cables laid by other States and passing through its waters without making landfall." The term "laid by other States" refers not only to cables laid by

49 See *Flores v Southern Peru Copper*, 343 F.3d 140 (2nd Cir. 2003) (United States) for general discussion on this point.

50 Article 21(1)(c), UNCLOS.

States but to those laid by their nationals.[51] The archipelagic State "shall permit the maintenance and replacement of such cables upon receiving due notice of their location and the intention to repair or replace them."[52] Given the passage of time, this provision has little practical utility since cables existing at the time UNCLOS entered into force have likely been retired. New cables that plan to transit archipelagic waters should obtain permission of the archipelagic State.

2 The Exclusive Economic Zone and Continental Shelf

Beyond territorial seas lie the EEZ and legal as opposed to geological continental shelf where coastal States enjoy certain sovereign rights over the exploration and exploitation of natural resources but where other States enjoy the rights of navigation and the freedom to lay and maintain submarine cables. The EEZ may extend up to 200 nautical miles (370 km) from the coastal baseline. The continental shelf depending upon qualifying geological criteria under UNCLOS may be recognized beyond the 200 nautical mile boundary up to a maximum 350 nautical miles (648 km).[53] The relevant UNCLOS articles on continental shelf and in the EEZ are Articles 56, 58, 78, and 79.

Paragraph 2 of Article 78 reiterates a consistent UNCLOS principle that coastal States must recognize the rights and freedoms of other States that are provided for in the Convention. It emphasizes that, in the exercise of its rights over the continental shelf, a coastal State must not infringe or cause unjustifiable interference with navigation and other rights and freedoms of other States as provided in the Convention, and the categoric character of this obligation is emphasized by the use of the words "***must not***."[54] The reference to "other rights and freedoms of other States" includes rights regarding submarine cables.[55]

The rights and freedoms of submarine cables on the continental shelf are set out in Article 79. The ordinary meaning of Article 79 is that repair or other

51 M.H. Nordquist (ed-in-chief), *United Nations Convention on the Law of the Sea 1982: A Commentary*, Vol. II (Martinus Nijhoff: Dordrecht/Boston/London, 1993), at p. 474 [51.7(i)]. Hereinafter *Virginia Commentary*.

52 Article 51(2), UNCLOS.

53 Article 76, UNCLOS.

54 *Virginia Commentary* (n. 51), Vol. II, at p. 906 [78.8(c)] (emphasis added).

55 Ibid., at p. 907 [78.8(d)].

permits[56] or taxes[57] or fees[58] imposed on international telecommunications and HVDC cables by coastal States outside of their territorial seas are not authorized under UNCLOS. Paragraph 1 of Article 79 affirms, in a form appropriate for the continental shelf, the provisions of Article 87(1)(c) [Freedom of the high seas][59] by which the freedom of the high seas enjoyed by all States includes the freedom to lay submarine cables.[60]

Paragraph 2 of Article 79 further expands the freedom of States to carry out "maintenance" of existing cables.[61] Maintenance is understood to include

---

56 Following a diplomatic protest by France, the US Army Corps of Engineers (USACE) withdrew its objection to the laying without a permit of the "AMX-1," international submarine cable 135 nautical miles off the coast of Puerto Rico in the United States EEZ. In an email dated 24 October 2013 from Mr. Noel Méndez (USACE) to Emmanuel Danjou, it was stated: "I have been asked to advise you that a decision has been reached concerning the US Army Corps of Engineering regulatory responsibility over cable laying activities on the seabed, specifically with the Exclusive Economic Zone (EEZ) under the Outer Continental Shelf Act (43 U.S.C. 1333(e)). We have concluded that we will limit our regulatory authority up to but not exceeding 12 nautical miles seaward."; confirmed also by email dated 2 March 2015 from Mr. Lance Wood (USACE) to Roy Carryer.

57 Supreme Court (Contentious-Administrative Division, 5th Chamber) Ruling of 16 June 2008, JUR 2008/211246 (*Telefónica de España S.A. v Ministry of the Environment*) holding that the imposition of taxes or fees outside of the territorial sea by the Spanish Government on a submarine cable violated UNCLOS. (English translation available upon request from the ICPC legal data base: www.iscpc.org).

58 An attempt by Malta to access a per kilometre fee on an international submarine cable that transited its continental shelf claim but never entered its territory or territorial sea was unsuccessful in the face of diplomatic challenges by the United Kingdom that the fee was inconsistent with UNCLOS. See D.R. Burnett, R.C. Beckman and T.M. Davenport (eds), *Submarine Cables: the Handbook of Law and Policy* (Brill/Martinus Nijhoff: Leiden/Boston, 2014), hereinafter: *Submarine Cables Handbook*, at p. 150 (Chapter 5, by K. Ford-Ramsden and T. Davenport, 'The Manufacture and Laying of Submarine Cables', 123–153).

59 In accordance with Article 87(1): '[t]he high seas are open to all States, whether coastal or land-locked. Freedom of the high seas is exercised under the conditions laid down by this Convention and by other rules of international law. It comprises, *inter alia*, both for coastal and land-locked States: (a) freedom of navigation [and] (c) freedom to lay submarine cables and pipelines, subject to Part VI [Continental Shelf]. 'Freedom of [n]avigation will include ... the movement and stationing of ships to lay and maintain submarine cables'; see Chapter 17, 'Navigation', in R.J. Dupuy and D. Vignes (eds), *A Handbook on the New Law of the Sea*, Vol. 2 (Martinus Nijhoff: Dordrecht /Boston/Lancaster, 1991), at p. 845. Hereinafter *Dupuy-Vignes Handbook*.

60 *Virginia Commentary* (n. 51), Vol. II, at p. 915 [79.8(a)].

61 Ibid. [79.8(b)]; 'laying' refers to new cables, while the term 'maintenance' relates to both new and existing cables. 'Repair' is an obvious part of maintenance of a submarine cable system.

repair. Paragraph 2 goes on to require that coastal States "may not impede the laying or maintenance of such cables." The only qualification of this mandate is in connection with the coastal States taking "reasonable measures for the exploration and exploitation of its natural resources"[62] and the "prevention, reduction and pollution from pipelines".[63] Telecommunication cables are not in any way involved with the exploration and exploitation of natural resources, and injury to a cable does not cause pollution.[64] The same is true for modern international HVDC power cables which use impregnated paper or polyethylene or non-oil based plastic insulation (i.e., XPLE, cross-linked polyethylene).[65] Therefore, a coastal State has no basis for its impeding the maintenance of these cable systems on the continental shelf by imposing permits, delays, taxes, fees, custom duties or guard boat requirements.[66] Nor can coastal States require through permit conditions that cable owners or cable ship operators

62 Article 77 defines natural resources as limited to mineral and non-living resources of the seabed and subsoil and living organisms of the sedentary species.

63 Article 79 reflects the careful compromise of the States which negotiated the text at UNCLOS III. China proposed that 'the delineation of the course for laying cables and pipelines in the seabed of the economic zone is subject to the consent of the coastal State'; see *Report of the Committee on the Peaceful Uses of the Sea-Bed and the Ocean Floor beyond the Limits of National Jurisdiction*, Vol. V, General Assembly, Official Records, Twenty-eighth Session, Supplement No. 21 (A/9021). A subsequent proposal by Denmark was adopted, however, which limited the coastal State's power to pipelines *only*, in recognition of the fact that a ruptured pipeline could result in pollution while a ruptured cable has no such result. See *Virginia Commentary* (n. 51), Vol. II, at p. 914 [79.7, n. 8]; also Chapter 18, 'Submarine Cables and Pipelines', in *Dupuy-Vignes Handbook* (n. 59), Vol. 2, at p. 985 [n. 37].

64 L. Carter, D. Burnett, S. Drew, G. Marle, L. Hagadorn, D. Bartlett-McNeil and N. Irvine, *Submarine Cables and the Oceans—Connecting the World*, UNEP–ECMC Biodiversity Series No. 31, 2009. This report compiles and analyses the environmental experience with cables in the marine environment since submarine cables were introduced into the ocean in 1850 and underscores the neutral to very minor effect a modern fibre-optic cable has on the marine environment.

65 Oil based insulation for international submarine power cables connecting States was generally phased out in the early 1990's in favour of non-polluting polyethylene, ethylene-propylene rubber or other superior forms of plastic insulation. 'About Power Cables', ICPC website, www.iscpc.org at Publications (2012); see M. Eccles, J. Ferencz and D. Burnett, 'Submarine Power Cables', in *Submarine Cables Handbook* (n. 58), 301–322.

66 Information provided by the United Nations on this issue, as set out by its Division for Ocean Affairs and the Law of the Sea (DOALOS), is instructive. It says: 'beyond the outer limits of the 12 nm territorial sea, *the coastal State may not* (*and should not*) *impede the laying or maintenance of cables*, even though the delineation of the course for the laying of such pipelines [i.e., *not cables*] on the continental shelf is subject to its consent' (emphasis added); see DOALOS website, 'Frequently Asked Questions', response to Question #7,

pay fishermen or other seabed users so they will not interfere with cable laying and repair-a requirement to which they are bound as a matter of law.[67]

Paragraph 4 involves two separate considerations: first, the right of the coastal State to establish conditions for cables or pipelines entering its territory or territorial sea; and second, the jurisdiction over cables that are constructed or used in connection with natural resources, artificial islands, installations, and structures under its jurisdiction. The second item does not apply to international telecommunication or State-to-State HVDC power cables. It would apply to a fibre-optic or power cable used as shore links to off-shore wind farms, tidal current generators, or oil and gas platforms. With respect to the first item, this is a direct function of the coastal State's sovereignty over its territory and territorial sea. Repairs beyond 12 nautical miles (22.2 km) are entirely outside these areas. Since paragraph 3 allows coastal States the right to delineate the route of pipelines and not cables, coastal States have no authority to require permits or impose conditions that infringe and cause unjustifiable interference with a cable owner's freedom to lay or carry out maintenance (or repair) of its cables.[68]

---

at: http://www.un.org/depts/los/LEGISLATIONANDTREATIES/frequently_asked_questions.htm (last accessed 1 March 2017).

67 *Ninety-Four Consortium Cable Owners v Eleven Named French Fishermen* (http://www.iscpc.org/members/Legal/ Precedence_Cases/Boulogne_Judgement_Iss_1.pdf), Tribunal de Grande Instance de Boulogne Sur Mer (1st Chamber) 28 August 2009, [File No. 06/00229 DG/LM]. Judgment in favour of consortium of cable owners against 11 French fishermen for damages caused by interference in cable repair ship operations by French fishing vessels. The French court found that the actions of the French fishermen to extract financial payments to allow the cable repair ships to repair a cable fault violated Articles R46 and R47 of the French Civil Code (CDFE), which require fishing vessels to keep all of their equipment and nets at least one nautical mile from the vessel repairing an underwater cable. (Note, this French domestic stature implements Article 5 of the 1884 Cables Convention.) The argument by the French fishermen that the cable was laid in French territorial seas without legal authorization was rejected. Instead, the court found that 'the measures taken against [the two cable ships] were part of a concerted attempt to obstruct the operation of underwater cables in fishing areas in return for financial compensation'. The court further found 'that each [fishing vessel] in question individually contributed to the incorrect interception of the two cable ships and sailors, as part of this action, played a part in an act of personal, even concerted misconduct that gave rise to damages'. Damages were awarded against the 11 named fishermen with interest for the delay of several days in carrying out the repairs. But damages do not remove the disruption cause by leaving vital international cables broken for weeks.

68 Article 79(2), UNCLOS.

Paragraph 5 underscores this point with its express provision that the "possibilities of repairing existing cables ... shall not be prejudiced." This provision is specifically cross-referenced in Article 112, which applies this limitation on coastal States to all cables on the high seas.

Paragraphs 2 and 4 reserve to the coastal State the power to regulate submarine cables used in connection with the exploration or exploitation of natural resources. Accordingly, cables used to connect oil and gas platforms or alternative energy facilities are subject to coastal State approval. Coastal State approval in the EEZ applies to cables used in connection with offshore wind, current, wave and tidal applications that generate power.[69] Similarly, cables used in connection with cabled observatories and other scientific purposes in the EEZ are subject to coastal State permission under the regime regulating marine scientific research.[70]

An important but deliberately vague obligation in paragraph 5 is that of "due regard."[71] Due regard is an obligation present in the EEZ, continental shelf and high seas regimes.[72] One case to address the meaning of "due regard" in an EEZ context held that it is driven by the particular circumstances involved, but includes elements of notice and meaningful consultation between the States involved.[73] Due regard arises in the context of competing uses by the coastal State under Article 56 and other States under Article 58. "The significance of this provision is that it balances the rights, jurisdiction and duties of the coastal State with the rights and duties of other States in the EEZ."[74] Coastal States often misunderstand or ignore this obligation when considering conflicts that arise between cables used in connection with natural resources, wind, and currents and marine scientific research, which they control, and those international cables that are used for telecommunications or shore generated power transfer between States where coastal State powers are very limited.

Article 58 addresses the rights and obligations of States regarding submarine cables in the EEZ. Paragraph 1 is explicit that the freedoms listed in Article 87, including navigation, which includes the stationing and movement

69 Article 56(1)(a), UNCLOS.

70 Article 56(1)(b)(i)(ii), UNCLOS.

71 See *Virginia Commentary* (n. 51), Vol. II, at p. 565 [58.10(e)].

72 Articles 56(2), 58(3), 79(5) and 87(2), UNCLOS.

73 *In the Matter of the Chagos Marine Protection Area Arbitration*, before an Arbitral Tribunal Constituted under Annex VII UNCLOS, between the Republic of Mauritius and the United Kingdom of Great Britain and Northern Island, Permanent Court of Arbitration, Award of 18 March 2015, para. 519.

74 See *Virginia Commentary* (n. 51), Vol. II, at p. 543 [56.11(f)].

of cable repair ships,[75] and "the laying of submarine cables ... and other internationally lawful uses of the sea related to these freedoms, such as those associated with the operation of ... submarine cables," are recognized in the EEZ.[76] Maintenance and repair by cable ships and cable route surveys incident to laying cables are lawful operations associated with the operation of cables.[77]

Paragraph 2 provides for the application in the EEZ of the high seas Articles 112 through 115 and other pertinent rules of international law in so far as they are not incompatible with UNCLOS, Part V (EEZ).

Paragraph 3 of Article 58 limits the ability of the coastal State to enforce laws and regulations in the EEZ to those measures that are not incompatible with Articles 58(1) and (2) as described above. The issue of compatibility is in essence a competition of those activities in the EEZ that are free on the high seas (i.e., cables) and those for which the coastal State enjoys sovereign rights (i.e., EEZ fishing) or exclusive jurisdiction (e.g., marine scientific research). A reasonable interpretation is that priority should be given to the activity explicitly recognized as free, such as cables, over an activity not explicitly recognized.[78] This is consistent with the object of UNCLOS to "facilitate international communication"[79] and that a coastal State may not exceed its powers in the EEZ.[80]

3 The High Seas

Articles 87(1)(c) and 112 through 115 apply on the high seas. They also apply on the continental shelf, and in the EEZ except as modified in those regimes. They are important in modern times because they effectively resolve conflicts among seabed users. As noted in Article 87(2), "these freedoms shall be exercised by all States with *due regard* for the interests of other States in their

75 See Chapter 17, in *Dupuy-Vignes Handbook* (n. 59), Vol. 2, at p. 872.

76 Ibid.

77 In contrast, the UNCLOS Article 87 freedoms of fishing and marine scientific research are not included. See *Virginia Commentary* (n. 51), Vol. II, at p. 564 [58.10(a)].

78 See Chapter 17, in *Dupuy-Vignes Handbook* (n. 59), Vol. 2, at pp. 872–873. In this chapter, Judge Treves recognizes two caveats to this priority interpretation. The first is the protection of human life. In normal submarine cable repairs protection of human life is not involved. The second is a determination that the two activities cannot co-exist. The two activities cannot co-exist as the ability to maintain submarine cables is significantly compromised to the point that cable owners suffer damages and its communications are impeded and placed at risk.

79 Preamble, UNCLOS.

80 See *Virginia Commentary* (n. 51), Vol. II, at p. 565 [58.10(e)].

exercise of the freedom of the high seas, and also with *due regard* for the rights under this Convention with respect to activities in the Area." (*emphasis added*)

Article 112 [Right to lay submarine cables and pipelines] provides for the freedom to lay submarine cables on the bed of the high seas beyond the continental shelf limited only by the caveat in paragraph 2 that, following Article 79(5), the laying shall occur with *due regard* to existing submarine cables and pipelines and in particular to the possibility of their repair.

Article 113 [Breaking or injury of a submarine cable or pipeline] deals with wilful or culpably negligent actions by mariners and fishermen that damage cables from contact with anchors, nets or other fishing gear. It requires States to enact national laws to provide for criminal sanctions against wrongdoers and vessels that injure international cables wilfully or by culpable negligence. Compliance is poor.

A few States such as Australia, New Zealand, Uruguay, and Colombia have modern and extremely effective modelled laws that generally comply with UNCLOS.[81] In these countries since legislation was enacted proactive monitoring of cables and associated protection zones together with effective enforcement of domestic laws have essentially reduced cable faults to near zero. The key in these cases is not only having modern legislation, but credible enforcement which together act as a deterrent to conduct by vessels harmful to cables. Where, like China, a State has modern legislation on paper, but never enforces its laws against fishing vessels, high fault rates are a matter of record.[82] Other States like the United States and the United Kingdom have telegraph era statutes based on the 1884 Cable Convention that are historical relics with little practical utility.[83] The majority of State parties to UNCLOS have *no* domestic laws that comply with the Convention.

Besides the treaty or statutory remedies, damages for injury to submarine cables are typically dealt with by civil suits in traditional admiralty courts under the general maritime law where the offending vessel is subject to

81 Protection of Submarine Cables and Other Measures Act of 2005 [Australia]; Submarine Cable and Pipeline Protection Act of 1996 [New Zealand]; Penal Code, Articles 128, 148, and 217 (2011) [Uruguay]; Resolution 204, Director General Marine (2012) [Colombia].

82 See Annex (below), that records for China in the years 2008–2015, an average number of about 26 cable faults per year, the highest of any State.

83 Submarine Cable Protection Act of 1887 [United States]; Submarine Telegraph Act of 1885 [United Kingdom].

arrest.[84] These remedies are especially popular in situations where a State has no legislation or obsolete legislation, but fair and efficient admiralty courts.

Article 115 [Indemnity for loss incurred in avoiding injury to a submarine cable or pipeline] complements Article 113 by providing an indemnity from the cable owner to any mariner or fishermen who, despite taking reasonably precautionary measures beforehand, sacrifices his anchor, net or fishing gear to avoid injury to a cable.[85] The indemnity, however, is limited to the sacrificed gear or anchor and does not include lost profits or catch.[86] These two articles reflect the very successful balancing and practical compromise of the competing uses of submarine cables on the one hand and fishing and navigation on the other.[87]

Article 114 [Breaking or injury by owners of a submarine cable or pipeline of another submarine cable or pipeline] addresses the indemnity for repair costs paid by the cable or pipeline owner for injury to prior laid cables or pipelines damaged during the laying of a subsequent cable or pipeline.[88] The indemnity in the case of a pipeline is limited to the actual repair costs and does not include compensation for any financial losses of the owner or the contents of a broken pipeline.[89] Again, this article illustrates a practical and common sense approach to the conflict that would otherwise arise with successive laying of

84 *The Government of the Netherlands, Post Office v G'T Manneteje-Van Dam [Fishing Cutter GO 4]*, File No. 325/78 (District Court Rotterdam, decision rendered 20 November 1978), aff'd *sub nom* G.'t Mannetje-Post Office, File No. 69 R/81 and File No. rb 325/78 (The Court at the Hague, Second Chamber, decision rendered 15 April 1983); M.M. Whiteman, *Digest of International Law*, Vol. 9 (US State Department: Washington DC, 1968), at 948 (Alex Pleven); *AT&T Corp. v Tyco Telecommunications, Inc.*, 255 F. Supp. 2d. 174 (S.D.N.Y. 2003); *American Telephone and Telegraph v M.V. Cape Fear*, 763 F. Supp. 97 (D.N.J. 1991), revised on other grounds, 967 F.2d. 864 (3rd Cir. 1992); Arbitration between Concert Global Network Services, Ltd., in its own capacity, and as co-maintenance authority of submarine cable system TAT-10, as Claimant and Tyco Telecommunications (U.S.) Inc. as Respondent, (Arb. New York, SMA 3779, 2002).

85 See *Virginia Commentary* (n. 51), Vol. III, at p. 277 [115.7].

86 Article 7, Cable Convention, and Renault 1884 (n. 18), at p. 13; and *Agincourt Steamship Company Ltd. v Eastern Extension, Australia and China Telegraph Company Ltd.* [1907] 2 K.B. 305.

87 The compromise reflected in Articles 113 and 115 is directly derived from Articles 2 and 7 of the 1884 Cable Convention and are widely followed as the custom and practice of the cable industry.

88 Arbitration between Concert Global Network Services, Ltd., in its own capacity, and as co-maintenance authority of submarine cable system TAT-10, as Claimant and Tyco Telecommunications (U.S.) Inc. as Respondent, (Arb. New York, SMA 3779, 2002).

89 See *Virginia Commentary* (n. 51), Vol. III, at p. 273 [114.7(b)].

cables and pipelines on the same seabed area.[90] Cable industry practices for cable crossings embrace this common sense approach.[91] With the exception of the Arabian Gulf where energy companies, often affiliated with coastal States, improperly demand one sided and onerous crossing agreements for pipeline crossings that violate UNCLOS, this article has been successfully implemented.

Finally, in the context of disputes of competing uses in the EEZ, upon the continental shelf, or on the high seas, it is of special importance to recognize that the laying and maintaining of submarine cables enjoys the highest level of protection under the UNCLOS dispute resolution provisions where such disputes with coastal States are subject to the mandatory requirements of these provisions.[92]

As the above summary demonstrates, the time proven submarine cable provisions in UNCLOS are a thoughtful balance of rights and obligations, which adequately provide the legal structure that has allowed the cable community to provide the amazing international internet system that is the critical international infrastructure of modern global society. Great care, careful thought, and evidence justifying the need and the risk of unintended consequences of changing this careful international law balance, are elementary.

III Submarine Cables and the Marine Environment Beyond the Limits of National Jurisdiction

Although the area beyond national jurisdiction (ABNJ) is remote and deep, there is still sufficient knowledge to provide a general overview of its environment. Satellites continually observe the temperature, height, currents and plankton content of the ocean surface as well as the topography of the ocean floor.[93] Back on Earth, various observatories constantly monitor global seismic activity, tsunamis and ocean currents.[94] Such information is complemented by

90 The compromise reflected in Article 114 is directly derived from Article 4 of the 1884 Cable Convention.

91 ICPC Recommendation No. 9A Telecommunication Cable and Oil Pipeline/Power Cable Crossing Criteria, available upon request www.iscpc.org.

92 Article 297(1) [Limitation of applicability of section 2].

93 NASA, 2017. *Ocean Colour Web*; *Data* at: https://oceancolor.gsfc.nasa.gov/, last accessed on 14 May, 2017; D.T. Sandwell, R.D. Müller, W.H.F. Smith, E. Garcia and R. Francis, 'New global marine gravity model from CryoSat-2 and Jason-1 reveals buried tectonic structure' (2014) 346 *Science* 65–67.

94 US Geological Survey, 2016. *Earthquakes*, at: http://earthquake.usgs.gov/earthquakes/; and Argo, 2016. *Argo Program*, at: http://www.argo.ucsd.edu/ (both last accessed on 6 December 2016).

a wealth of regional studies from ship-borne surveys. However, the biological world is less well covered. Global programmes such as the Census of Marine Life are advancing knowledge especially of the North Atlantic and Pacific oceans but are also exposing data-poor regions in the high Arctic and large parts of the Southern Hemisphere.[95]

A *Environmental Setting*

TABLE 2 *Environmental setting of submarine cables*

Oceans' area (km^2)	362 million km^2 or 71% of Earth's surface
Average ocean depth (m)	3688 m
ABNJ (km^2)	ca. 230 million km^2 or ca. 39% of Earth's surface (this is an approximation that reflects uncertainties of some EEZ and continental shelf boundaries)
Average depth (m)	>3688 m as the area usually excludes the 0 to 2000-3000 m-deep continental margin, which commonly resides within the EEZ and/or the continental shelf (Figure 1)

1 Depth and Shape of the Deep Ocean

The surface area beyond the limits of national jurisdiction is substantial as shown in Figure 1, which outlines the regions beyond EEZs.

95 Ocean Biogeographic Information System, 2016. *Where is what and what is where?* at: http://news.coml.org/descrip/imgs/OBIS_marbef_articl03.05.pdf (last accessed on 14 May 2017).

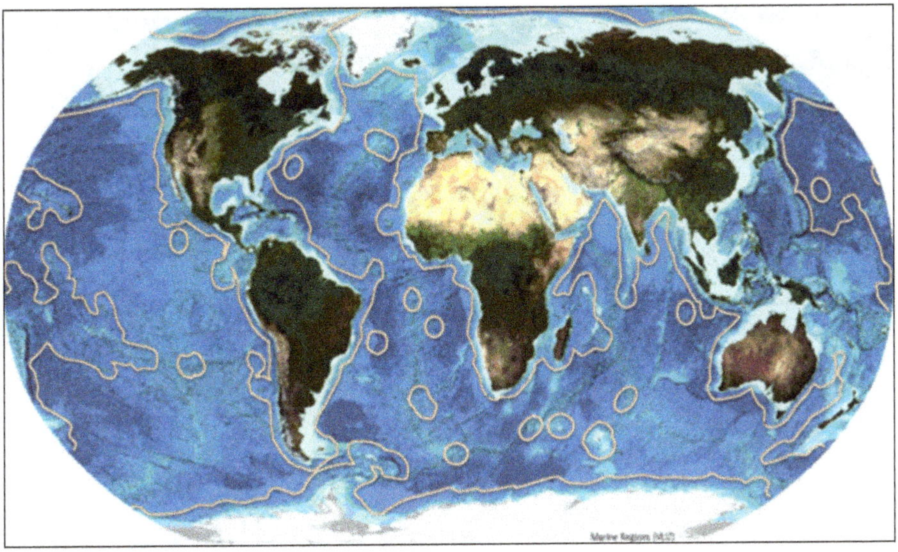

FIGURE 1 *A generalised chart of seaward boundaries of* EEZS *(pink lines) that includes a hypothetical 200 nautical mile limit (370 km) around Antarctica.*[96] *The* EEZ *encompasses the shallow geological continental shelf* (0–ca. *130 m depth), continental slope and rise* (ca. *130 m to 1500–3000 m) and, in regions such as the circum-Pacific rim, even abyssal depths. In contrast, the* ABNJ *is mostly the abyssal ocean plains and hills that typically extend below* ca. *3000 m depth. Mid-ocean ridges, such as the mid-Atlantic Ridge that pass through the* ABNJ, *are mainly deeper than* ca. *3000 m but can locally shallow to less than 1500 m.*
SOURCE: CHART COMPILED FROM INFORMATION IN MARINE REGIONS.ORG.[97]

A cable in the ABNJ typically traverses a diverse ocean floor comprised of abyssal plains and hills, mountainous ridges sometimes larger than their terrestrial counterparts, plateaux and innumerable submarine volcanoes or *seamounts* (for an illustration, see Figure 2).[98] There are also depressions. Trenches can extend to 6 km below the adjacent ocean floor. Channels can continue from submarine canyons, and wend across the ocean floor for hundreds to thousands of kilometres.

96 Marine Regions.org, 2017. *World maps* at: http://www.marineregions.org/maps.php?album=3264&pic=64930 (last accessed 14 May 2017).

97 Ibid.

98 Sandwell et al. (n. 93), at pp. 65–67.

FIGURE 2 *Although within the* EEZ *of the Mariana Islands, this digital terrain chart of the Northern Mariana Arc is relevant as an illustration. It exemplifies the topography associated with seamounts that dot the ocean floor of the* ABNJ. *Where possible, this topography is avoided by cable routes because of hazardous steep volcanic slopes, rough rocky topography, potential volcanic activity and locally intensified ocean currents. Seamounts may also support rich and diverse biological communities, which attract commercial fishing especially on the high seas.*
SOURCE: BILL CHADWICK, OREGON STATE UNIVERSITY AND NOAA PACIFIC MARINE ENVIRONMENTAL LABORATORY; HTTP://OCEANEXPLORER.NOAA.GOV/EXPLORATIONS/14FIRE/BACKGROUND/SEAMOUNTS/SEAMOUNTS.HTML (LAST ACCESSED ON 31 DECEMBER 2016).

2 Ocean Currents

The ocean floor of the ABNJ is subject to major currents that connect the Northern and Southern Hemispheres. Super cold, dense water is formed around Antarctica and spreads northwards in water depths greater than 2000 m (Figure 3). At the same time, surface currents in the North Atlantic Ocean, in particular the Gulf Stream, cool as they flow north, providing heat for Europe and the eastern seaboard of North America. As this water cools it

FIGURE 3 *Currents of the global ocean overturning circulation as viewed for the Pacific and Indian Oceans. Of relevance are the deep currents (blue arrows* ca. *>2000 m water depth) that also circulate over the seabed of the* ABNJ*. These flows can intensify against steep seabed slopes, such as those of seamounts, to reach speeds capable of moving sediment. This movement may fatigue or abrade cables.*
SOURCE: CSIRO HTTP://WWW.CMAR.CSIRO.AU/CURRENTS/ANIMATIONS.HTM (LAST ACCESSED ON 6 DECEMBER 2016).

too becomes dense and sinks to return south to Antarctica.[99] Normally these deep currents are slow *ca.* 5–10 cm/s (0.1–0.2 knots) but they can intensify against steep submarine topography especially that of continental margins, submarine ridges and seamounts.[100] Currents can accelerate to erode the

99 H. Stommel, 'The abyssal circulation' (1958) 5 *Deep-Sea Research* 80–82; J. Marshall and K. Speer, 'Closure of the meridional overturning circulation through Southern Ocean upwelling' (2012) 5 *Nature Geoscience* 171–179.

100 I.N. McCave and L. Carter, 'Sedimentation beneath the Deep Western Boundary Current off northern New Zealand' (1997) 44 *Deep-Sea Research* 1203–1237.

seabed in depths of 3000–4000 m and deeper.[101] This interaction with zones of strong relief can also form giant eddies, which in the case of the Gulf Stream disturb the seabed to around 5 km water depth.[102]

3 Natural Hazards

The risk of natural hazards is an important factor in cable route selection. The distribution and frequency of hazards vary with the regional geology, climate and oceanography. The most risk-prone regions are where tectonic plates collide; a phenomenon marked by extensive earthquake and volcanic activity, extreme terrestrial erosion and disproportionally large discharges of river sediment into the ocean. Taiwan, for example, is smaller than Iceland but produces about 2% of the sediment reaching the world ocean.[103] The most extensive region of tectonic plate collision is the rim of the Pacific Ocean, popularly known as the *Pacific Ring of Fire.* Lesser, but nonetheless important collisional zones also dominate the Caribbean, NE Indian Ocean and the Mediterranean regions.

The continental margins of such seismic regions are subject to submarine landslides that can transform into *turbidity currents.*[104] These turbulent sediment-laden flows can reach speeds of 68 km/hour and travel hundreds of kilometres commonly guided by submarine canyons and channels. And if cables lie in a current's path there is the threat of breaks.[105]

If a seamount is active, potential threats include lava flows, hot-water vents, earthquake- or volcanic-triggered landslides and debris flows. Furthermore, rugged rocky topography may suspend cables that can be damaged by fast currents. Extinct seamounts are a lesser risk, but the hazards of intensified currents and rough terrain remain (see Figure 2).

---

101 Ibid.

102 C.D. Hollister and I.N. McCave, 'Sedimentation under deep-sea storms' (1984) 309 *Nature* 220–225.

103 S.-J. Kao and J.D. Milliman, 'Water and sediment discharge from small mountainous rivers, Taiwan: The roles of lithology, episodic events, and human activities' (2008) 116 *Journal of Geology* 431–448.

104 P.J. Talling, C.K. Paull and D.J.W. Piper, 'How are subaqueous sediment density flows triggered, what is their internal structure and how does it evolve? Direct observations from monitoring of active flows' (2013) 125 *Earth-Science Reviews* 244–287.

105 L. Carter, R. Gavey, P.J. Talling and J.T. Liu, 'Insights into submarine geohazards from breaks in subsea telecommunication cables' (2014) 27(2) *Oceanography* 58–67, at: http://dx.doi.org/10.5670/oceanog.2014.40.

The impact of icebergs, sea ice, storm surges and tsunami are felt primarily at the coast and continental shelf and have minimal effect on the ABNJ,[106] unless large turbidity currents are formed that are capable of reaching the area.

B *Cables and Their Interaction with the ABNJ Physical Environment*

1 Cable Physical and Chemical Presence

Given that the average depth within the ABNJ exceeds 3688 m, cables are laid directly on top of the seabed. This reflects an absence of activities known to break cables—specifically ships' anchoring and bottom trawl fishing.[107] Accordingly, there is no requirement for protective burial below the seabed thus minimising any disturbance to the benthic environment. Furthermore, for water depths exceeding *ca.* 2000 m, telecommunications cables are typically 17–22 mm diameter, hence their physical foot print is small.[108] As 80% of the ocean is deeper than 2000 m then a similar percentage of cables are of the small-diameter "lightweight" design (Figure 4).

Certainly, water depths in the ABNJ favour the deployment of lightweight cable, i.e., one comprised of a high-grade, marine-quality polyethylene tube with a core of steel wire for strength, a copper conductor to power acoustical repeaters and glass fibres for communications.[109] There is no need for protective armour, and anti-fouling agents are not used as a matter of practice.[110] In that context, lightweight cables are chemically inert.[111]

Such is their chemical stability that decommissioned fibre-optic cables and their coaxial predecessors, have become targets of an expanding recycling

106 L. Carter, 'Submarine Cables and Natural Hazards', in *Submarine Cables Handbook* (n. 58), 237–254.

107 M.E. Kordahi, S. Shapiro, and G. Lucas, 'Trends in submarine cable system faults', *Proceedings SubOptic 2007*, Baltimore, 4 pp, at: http://www.suboptic.org/document/presentation-trends-in-submarine-cable-system-faults/, last accessed 6 December 2016.

108 Carter, 'Submarine Cables and Natural Hazards', in *Submarine Cables Handbook* (n. 58), pp. 237–254. See also Carter et al. (n. 64).

109 Ibid.

110 Emu Ltd, *Subsea cable decommissioning: A limited environmental appraisal.* Report no 04/J/01/06/ 0648/0415 (2004). Open file report available from European Subsea Cables Association, at: http://www.escaeu.org/

111 K. Collins, *Isle of Man Cable Study—preliminary material environmental impact studies*, Preliminary Report, University of Southampton, 2007 (prepared for BT, Global Marine Systems Ltd and Department of Environment, Food and Agriculture Isle of Man). NESDI, 'Studying the impact of seafloor cables on the marine environment' (Spring 2014) *Currents* 6–21, at: http://greenfleet.dodlive.mil/currents-magazine/currents-magazine-2014/currents-spring-2014/ (last accessed on 6 December 2016).

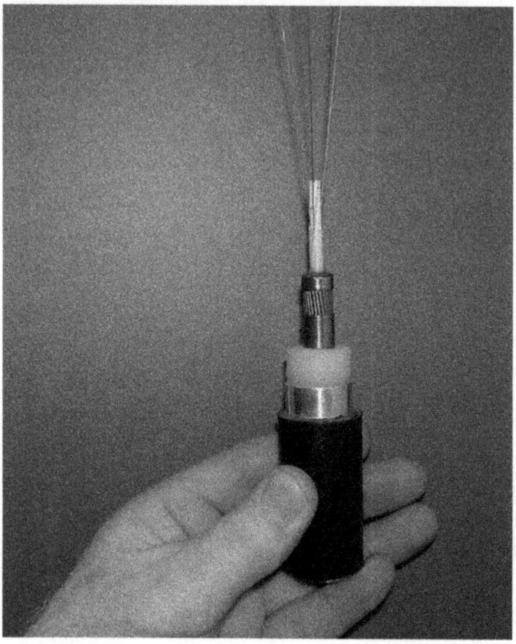

FIGURE 4
A lightweight fibre-optic cable with (from outside to core) black and white polyethylene sheath, power conductor (copper), steel wire strength member, glass fibres and plastic support sheath (white).
SOURCE: L. CARTER.

industry even though these cables may have lain on the seabed for three decades and longer.[112]

2 Cables and Natural Hazards

Taking a broad view, between 150 and 200 cable faults occur each year worldwide. Between 60-70% of those faults are caused by human activities, especially fishing and shipping.[113] Thus it is unsurprising that most faults occur on the geological continental shelf in depths less than *ca.* 200 m[114] as well shown by the global distribution of 2,162 faults for 1959-2008 (Figure 5).[115] Faults

112 See ICPC Recommendation No 1, 11 May 2011, available upon request at www.iscpc.org. Also D. Burnett, 'Out-of-Service Submarine Cables', in *Submarine Cables Handbook* (n. 58), 213–222.

113 M.E. Kordahi and S. Shapiro, 'Worldwide trends in submarine cable systems', *Proceedings SubOptic 2004*, Monaco, 2004, paper 3 pp, at: http://www.suboptic.org go to Resources/ Conference Archives/ SubOptic 2004, (last accessed on 14 May 2017); M.E. Kordahi, S. Shapiro and G. Lucas, 'Trends in submarine cable system faults', *Proceedings SubOptic 2007*, Baltimore, 2007, 4 pp, at: http://www.suboptic.org l go to Resources/ Conference Archives/ SubOptic 2007 (last accessed on 14 May, 2017).

114 Kordahi et al., 'Trends in submarine cable system faults' (n. 113).

115 S. Drew, 'Submarine cables and other activities', in Carter et al. (n. 64).

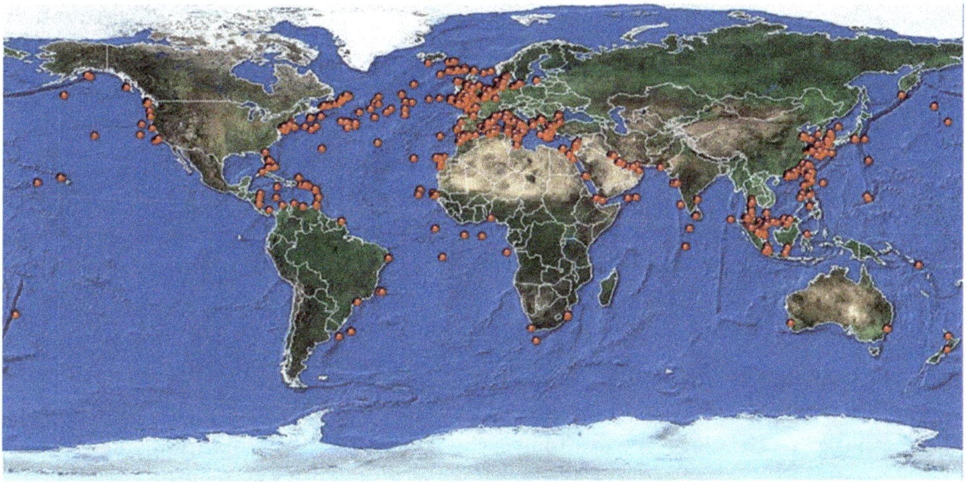

FIGURE 5 *This distribution of 2,162 cable faults was recorded between 1959 and 2008, a period that covers the last of the telegraphic cables, all coaxial cables and modern fibre-optic systems, the latter coming to prominence in the late 1980s. Most faults occurred in the shallow (< 200 m deep) seas around Europe, SE Asia and eastern seaboard of North America and reflect the intense fishing and shipping activity of those regions. Faults across the North Atlantic Ocean in part result from the failure of old telegraphic cables due to abrasion and component failure. For the rest of the ABNJ, cable faults are presently on average 4 per annum worldwide.*[116]
SOURCE: TE SUBCOM, WHICH GRANTED PERMISSION FOR USE.

due to failure of cable components are ≤ 5% whereas faults caused by natural hazards such as submarine landslides, are ≤ 10% of all faults, and tend to occur in water depths ≥ 1200 m.[117]

The most common natural causes of faults in the deep ocean are submarine landslides and associated turbidity currents.[118] Today, key areas of concern are

116 Palmer-Felgate, A., N. Irvine, S. Ratcliffe and S. Sui Bah, 'Marine Maintenance in the Zones—a Global Comparison of Repair Commencement Times', 2013, paper available at: http://www.suboptic.org/document/marine-maintenance-in-the-zones-a-global-comparison-of-repair-commencement-times/. (New data added to update record to 2015; see Annex, below.).

117 Kordahi and Shapiro, 'Worldwide trends in submarine cable systems' (n. 113).

118 Carter et al., 'Insights into submarine geohazards' (n. 105), 58–67; Carter et al. (n. 64); A. Cattaneo, N. Babonneau, G. Ratzov, G. Dan-Unterseh, K. Yelles, R. Bracène, B. Mercier de Lèpinay, A. Boudiaf and J. Déverchère, 'Searching for the seafloor signature of the 21 May 2003 Boumerdès earthquake offshore central Algeria' (2012) 12 *Natural Hazards Earth System Science* 2159–2172.

offshore Taiwan[119] and Algeria[120] where earthquakes in 2006 and 2003 caused 22 and 29 cable breaks, respectively. Damaging turbidity currents are also generated by cyclonic winds and rainfall through rivers discharging mud-laden floodwaters that dive to and along the seabed. Likewise, storm waves disturb seabed sediment that also may evolve into a turbidity current.[121]

While zones of tectonic plate collision produce cable-damaging turbidity currents on an annual basis, their impact on cables in the ABNJ is likely to be modest because (a) the main areas of landslides and turbidity currents are the geological continental shelves and slopes, which usually occur within EEZs (Figures 1 and 2); and (b) any turbidity current leaving an EEZ is likely to move on to a near-flat floor of the ABNJ where current speed reduces to a level where it is no longer a threat to a cable.[122] However, there are local departures from that generalisation. One example is the 1929 Grand Banks earthquake, which produced a turbidity current that travelled 800 km to break telegraph cables seaward of the Canadian EEZ.[123]

Where possible, cable route planners avoid zones of active landslides and turbidity currents such as submarine canyons and channels, but this is not always possible. The circum-Pacific Rim, for example, accommodates major cities such as Manila, Tokyo, San Francisco, Los Angeles, Santiago and others. And these metropolises rely heavily on the services provided by submarine cables, which therefore must traverse the hazardous Pacific Rim. Plotting the least hazardous route requires up-to-date knowledge of deep-ocean hazards as demonstrated for the Strait of Luzon. There, at least 17 fibre-optic cables cross a highly active submarine canyon and adjoining Manila Trench. These cables connect Southeast Asia to the rest of the world (Figure 6). New research is suggesting that future cable routes could cross deeper parts of the Manila Trench where turbidity currents slow down and are less destructive.[124]

---

119 Carter et al., 'Insights into submarine geohazards' (n. 105), 58–67.

120 Cattaneo et al., 'Searching for the seafloor signature of the 21 May 2003', (n. 118), 2159–2172.

121 Carter et al., 'Insights into submarine geohazards' (n. 105), 58–67; A.T. Dengler, P. Wilde, E.K. Noda and W.R. Normark, 'Turbidity currents generated by Hurricane IWA' (1984) 4 *Geomarine Letters* 5–11.

122 R. Gavey, L. Carter, J. Liu, P. Talling, R. Hsu, E. Pope and G. Evans, 'Frequent sediment density flows during 2006 to 2015, triggered by competing seismic and weather events: Observations from subsea cable breaks off southern Taiwan' (2016) *Marine Geology*, at: doi:10.1016/j.margeo.2016.06.001.

123 B.C. Heezen and M. Ewing, 'Turbidity currents and submarine slumps, and the 1929 Grand Banks earthquake' (1952) 250 *American Journal of Science* 849–873.

124 Gavey et al., 'Frequent sediment density flows' (n. 122).

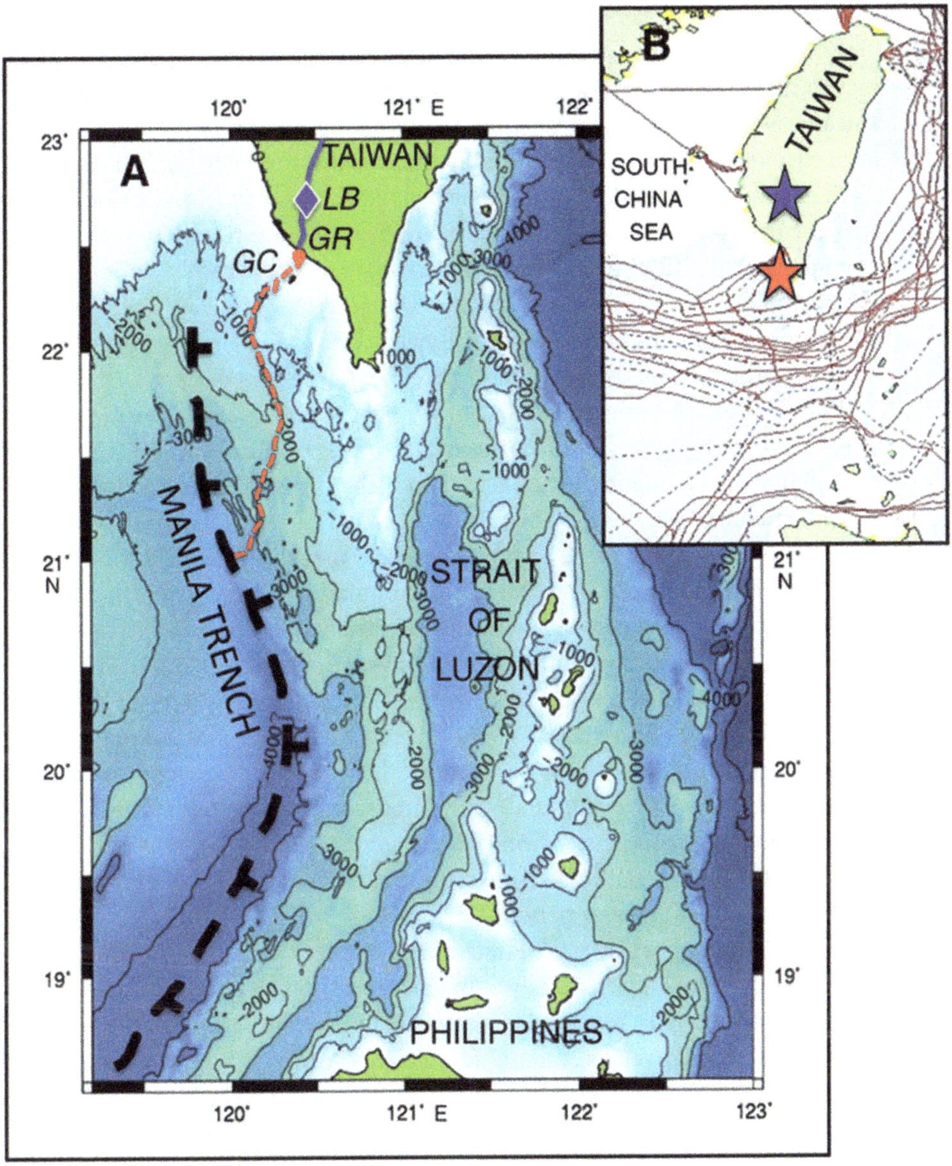

FIGURE 6 *The Strait of Luzon (A) contains at least 17 fibre-optic cables (B) and most cross the geologically active Gaoping Submarine Canyon (GC; course in red) and Manila Trench. The combination of frequent earthquakes and typhoons cause landslides and turbidity currents.*[125]

NOTE: THE STARS IN INSET B ARE THE EPICENTRES OF EARTHQUAKES THAT CAUSED MULTIPLE CABLE BREAKS IN 2006 (RED) AND 2010 (BLUE).

125 Ibid.

In the same vein, cable route planners avoid volcanically active seamounts and ocean ridges. This approach is successful as there are few cable faults directly attributed to lava flows, debris flows, hot-water vents and other types of volcanic activity. The main threat comes from the rough volcanic terrain and local currents that intensify in rocky gaps and against the steep sides of extinct and active seamounts. If a cable is suspended across a rough rocky terrain, strong currents may cause the cable to swing or vibrate resulting in fatigue and failure at the suspension points. Alternatively, a cable can be abraded by moving sand.

Risks from climate change are probably low due to the large depths of the ABNJ. Water depth tends to dampen more obvious effects of climate change that are witnessed in the upper ocean, namely strengthening storms, rising sea level and changing ocean currents and waves.[126] However, there is the potential of strengthening winds to enhance the power and/or frequency of bottom-stirring ocean eddies such as those associated with the Gulf Stream, but this hypothesis has yet to be substantiated. There is also a suggestion that stronger and wetter tropical cyclones will form large floods capable of transforming into turbidity currents with sufficient power to break cables as seen off Taiwan.[127] Even so, any climate-forced Taiwanese turbidity current is likely to have a minimal effect in the ABNJ, as the Area is located well to the east of Taiwan and any turbidity current would be trapped by the intervening submarine topography. Furthermore, studies show turbidity currents weaken along the near-flat floor of Manila Trench thus reducing the threat to cables.[128] Finally, climate change may affect deep-ocean currents but changes in flow speed are likely to be modest, as inferred from reconstructions of current behaviour in past major warm periods.[129]

Overall, the number of cable faults from all causes, including natural hazards, is small in the ABNJ. An analysis undertaken of cables in the Sargasso Sea, for example, revealed a total of three cable repairs for the period 2008–2015.[130] Those faults appear related mainly to abrasion in the vicinity of a chain of

---

126 Intergovernmental Panel on Climate Change, *Climate Change 2013: The Physical Science Basis. Working Group I Contribution to the IPCC 5th Assessment Report—Changes to the Underlying Scientific/Technical Assessment* (IPCC-XXVI/Doc.4), at: http://www.ipcc.ch/report/ar5/wg1/#.Uq5tSxaWfHg, last accessed on 6 December 2016.

127 Gavey et al., 'Frequent sediment density flows' (n. 122).

128 Ibid.

129 I.R. Hall, I.N. McCave, N.J. Shackleton, G.P. Weldon, and S.E. Harris, 'Glacial intensification of deep Pacific inflow and ventilation' (2001) 412 *Nature* 809–812.

130 A.L. de Juvigny, T.M. Davenport, D.R. Burnett and D. Freestone, 'Submarine telecommunication cables in the Sargasso Sea' (2015) 30 *International Journal of Marine and Coastal Law* 371–378.

extinct seamounts. Worldwide, the annual rate of cable faults in the ABNJ for 2008–2015 was on average four faults per annum.[131] (See Annex, below.)

3 Cable Operations

The laying of a submarine cable is guided by a desk-top study followed by a ship-borne survey to identify a safe, commercially-viable and environmentally neutral route.[132] Substantial databases exist that provide useful initial insights into a new cable route. Nevertheless, an actual route survey is required to specify the detail necessary for a successful cable deployment. In the ABNJ, these surveys are normally limited to a single pass of multibeam sonar and continuous seismic profiler. Where possible, cables avoid obvious natural hazards, zones of biological significance and cultural heritage sites.[133] Such diversions may require additional mapping. Route surveys are an inseparable part of the freedom to lay cables.

As noted earlier, hazardous zones are not always possible to avoid. The first trans-oceanic cable, for example, took the most direct route between Ireland and Newfoundland. As a result it crossed the volcanic mid-Atlantic Ridge (MAR) that extends from the high Arctic to the Southern Ocean (Figure 1). Telegraphic cables, which operated from 1866 to the 1950s and early coaxial systems (1950s to 1960s) were subject to faults in the general region of the MAR. Those faults were largely attributed to abrasion and fatigue in areas of current-swept rocky seabed, and also component failure, bearing in mind components in older cables were less reliable than modern systems. Today, at least 21 fibre-optic cables traverse the MAR[134] with far greater reliability due to improved cable design, the development of accurate seabed mapping systems complimented by precise GPS navigation and advanced scientific knowledge.

Cables typically have a design life of 20 to 25 years,[135] but improvements in signal processing mean that existing cables can be upgraded to operate for 30 years or more. Whether the life is 20 or 30+ years, the deployment of a cable is a brief, infrequent event of minimal extent. This contrasts with more repetitive or more extensive long-duration activities such as commercial fishing, oil and gas exploitation and seabed mining.

131 Palmer-Felgate et al., 'Marine Maintenance in the Zones' (n. 117).

132 Carter, 'Submarine Cables and Natural Hazards', in *Submarine Cables Handbook* (n. 58), 237–254. Carter et al. (n. 64).

133 Ibid.

134 TeleGeography, 2016. *Submarine cable map*, at: http://www.submarinecablemap.com/.

135 Carter, 'Submarine Cables and Natural Hazards', in *Submarine Cables Handbook* (n. 58), 237–254. Carter et al. (n. 64).

Because of the lack of human activities like bottom trawling and anchoring in the ABNJ, repairs are rare (see Annex, below). Repairs involve the towing of a specialised grapnel that secures and cuts the cable.[136] The secured end is brought to the surface and tied to a surface buoy. The grapnel then recovers the other cable end, at which stage a new section of cable is inserted or *spliced* between the recovered cable ends. The repaired assembly is lowered to the ocean floor taking care it is laid under tension to ensure coils or loops are not formed. It is recognised that the ocean floor will be momentarily disturbed during recovery. A grapnel may disturb a swath of ocean floor up to 1 m wide and a few kilometres long,[137] but the precise nature of the disturbance will depend upon the seabed geology.

There is an argument that scars produced by grapnels may be short-lived. The few fault repairs (four per annum) in the ABNJ occur mainly in zones of current-swept sediment that abrades and fatigues cables.[138] Such seabed mobility also has the potential to naturally smooth out any scars.

As repairs are designed to last a cable's lifetime, repair operations *per se* are infrequent and briefly invasive over a limited area.

4 Cables and Marine Biota

Knowledge of the interaction of cables with the marine benthic biota is based mainly on studies undertaken on the continental margin where the abundance and diversity of marine species is higher than is presently known for the deep ocean. This situation reflects the distribution of nutrient- and plankton-rich surface waters that typically overlay continental margins but may also extend into the ABNJ via major currents such as the eddy-rich Gulf Stream (Figure 7).

Studies of organisms living on and in a seabed occupied by cables show no statistical difference in abundance and diversity compared to areas without

136 Dynamic Load Monitoring, UK, 2016. *Grapnel Data Sheet*, at: http://www.dlm-uk.com/wp-content/uploads/2014/07/Grapnels-datasheet.pdf (last accessed on 6 December 2016).

137 Ibid.

138 de Juvigny et al. (n. 130), 371–378. I.C. Wright, 'In situ modification of modern submarine hyaloclastic/pyroclastic deposits by oceanic currents: an example from the Southern Kermadec arc (SW Pacific)' (2001) 172 *Marine Geology* 287–307.

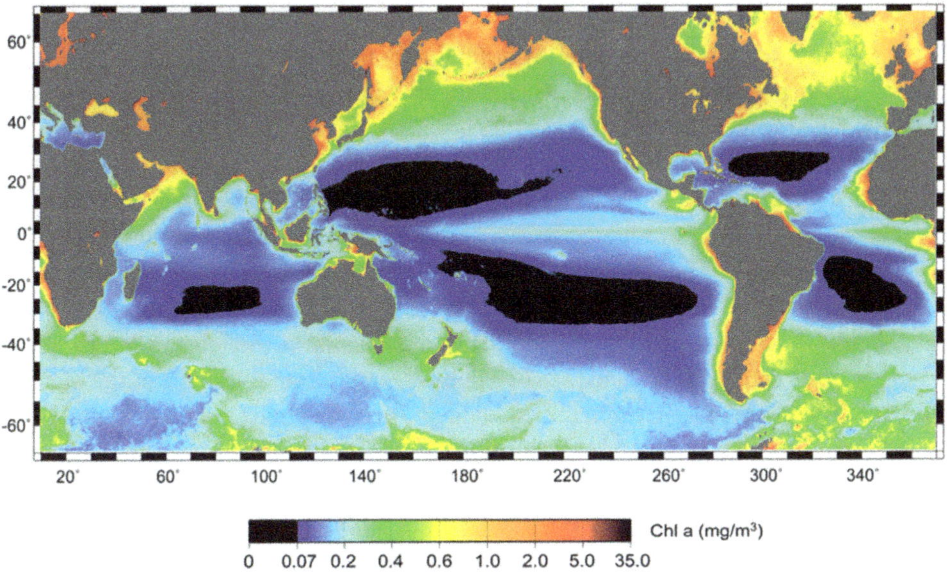

FIGURE 7 *Compiled satellite data showing regions of chlorophyll* a *in surface ocean waters. Chlorophyll* a *is a proxy for plant plankton, which underpins the marine food chain. Regions of highest production of plankton are red to yellow and the lowest are blue to black, which cover much of the* ABNJ. *Nevertheless, some major ocean surface currents may locally enhance chlorophyll* a *in the* ABNJ *such as in the North Atlantic Ocean.*
SOURCE: NOAA, HTTP://WWW.NOAANEWS.NOAA.GOV/STORIES2008/20080305_OCEANDESERT.HTML (LAST ACCESSED ON 22 DECEMBER 2016).

cables.[139] A recent study,[140] for example, is based on repeated seabed surveys made before (in 2004) and after (in 2007, 2010, 2015) deployment of a combined fibre-optic/power cable in Monterey Bay. Since the cable became operational, the abundance and distribution of skates and sharks, as well as animals larger than one millimetre inhabiting the seabed, showed few changes that

139 B.M. Grannis, *Impacts of mobile fishing gear and a buried fibre-optic cable on soft-sediment benthic community structure.* M.Sc. thesis, University of Maine, 2001, 100 pp; I. Kogan, C. Paull, L. Kuhnz, E. Burton, S. Von Thun, H.G. Greene and J. Barry, 'ATOC/Pioneer Seamount cable after 8 years on the seafloor: observations, environmental impact' (2006) 26 *Continental Shelf Research* 771–787.

140 L. Kuhnz et al., 'Potential impact of the Monterey Accelerated Research System (MARS) cable on the seabed and benthic faunal assemblages', 2015 *MARS Biological Survey Report*, at: https://www.mbari.org/wp-content/uploads/2016/02/MBARI-Potential-impacts-of-the-Monterey-Accelerated-Research-System-2015.pdf, (last accessed on 6 December 2016).

could be attributed to the cable. In essence, any measurable effect of the cable was overshadowed by the natural variability of the biota.

In the pre-1950s, whales, especially sperm whales, were observed entangled with old submarine telegraphic cables at the edge of the geological continental shelf.[141] However, the improved cable designs of coaxial and fibre-optic systems, plus better laying and repair procedures, were followed by a complete cessation of whale entanglements—a situation that continues to the present day (Figure 8).[142] In addition, most ABNJ water depths exceed 2000 m, which is the known diving limit of sperm whales.[143] Fish bites, including those of sharks, have damaged telecommunication cables from the telegraphic to fibre-optic cable eras.[144] Evidence of fish bites comes from the shape of the bite marks and the presence of teeth embedded in a cable's polyethylene sheath.[145] From 1901 to 1957 at least 28 telegraphic cables were damaged.[146] Between 1959 and 2006—a span that encompasses coaxial and fibre-optic systems—approximately 11 cables needed repair due to fish bite.[147] The most recent data covering 2007 to 2015 reveal no cable faults attributable to fish.[148] This

---

141 B.C. Heezen, 'Whales entangled in deep sea cables' (1957) 4 *Deep-Sea Research* 105–115.

142 M.P. Wood and L. Carter, 'Whale entanglements with submarine telecommunication cables' (2008) 33 *IEEE Journal of Oceanic Engineering* 445–460. See also: A. Coghlan, 'Hacker, the humpback whale who got entangled in an internet cable', *New Scientist*, Issue 3100, 19 November 2016. At the time of writing, a humpback whale was reported to be entangled with a cable of unknown type in Kaldfjord, Norway. The whale was freed, but the circumstances leading to its entanglement have yet to be identified. Until that information is to hand, action cannot be undertaken to ensure that the first entanglement since ca. 1959, is not repeated. Preliminary observations suggest the cable involved is an unburied, unusually yellow coloured, domestic fibre-optic cable laid presumably in Norwegian internal waters and/or its territorial sea within a fjord in a depth of about 170 metres. Information needed to better understand this rare incident include the reasons (1) why the cable was not buried in such shallow water, (2) why it was not armoured, (3) why there was apparent slack in the laying that allowed the cable to be lifted by the whale to the water surface, and (4) what desk top assessment and planning were made prior to the cable laying including information about whale activities in the location.

143 W.A. Watkins, M.A. Daher, K.M. Fristrup, T.J. Howald and G.N. Di Sciara, 'Sperm whales tagged with transponders and tracked underwater by sonar' (1993) 9 *Marine Mammal Science* 55–67.

144 Wood and Carter (n. 142).

145 L.J. Marra, 'Shark bite on the SL submarine light wave cable system: History, causes and resolution' (1989) 14 *IEEE Journal Oceanic Engineering* 230–237.

146 International Cable Protection Committee, 1988. Paper ICPC Plenary 1988.

147 S. Drew, 'Submarine cables and other activities', in Carter et al. (n. 64).

148 ICPC, 2015. Unpublished data from latest ICPC coordinated analysis, on file with the authors.

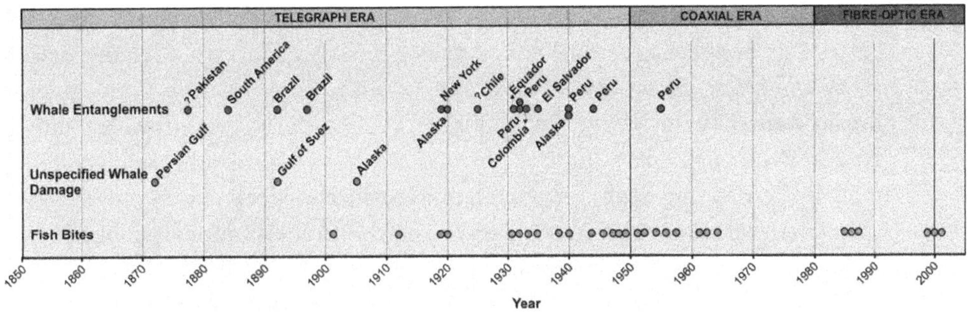

FIGURE 8 *The record of open-ocean whale entanglements, which ceased with the establishment of coaxial cables (1950s 1960s) and their fibre-optic successors (1980s to present day). Faults from fish bites have generally declined and have not been recorded since 2007.*[149]

marked decline in fish-related faults results mainly from improved cable design although reduced fish stocks may also be a factor.

5 Cables and Their Environmental Value

Since deployment of the first, fully operational trans-oceanic cable in 1866, cables have provided data and knowledge about the marine environment. Early telegraphic cables recovered from abyssal depths were sometimes encrusted with marine life or covered with sediment containing live organisms. These isolated discoveries came at a time of much debate in Victorian science circles as to whether life could exist in water depths greater than 500 fathoms (914 m). Cold temperatures, the lack of light and high pressures were thought by some scientists to be too extreme to support life.[150] To resolve the issue the first worldwide survey of the ocean, namely the HMS *Challenger* expedition, was instigated with a key aim to determine what life, if any, occupied the oceanic abyss.[151] The four-year venture between 1872 and1876 revolutionised mid-19th Century views of the ocean with the discovery of many new organisms at all ocean depths.

149 Wood and Carter (n. 142).

150 Natural History Museum, *Treasures of the Natural History Museum* (Natural History Museum: London, 2008), 256.

151 Challenger Society for Marine Science, 2016. *History_of_the_Challenger_Expedition* at: http://www.challenger-society.org.uk/History_of_the_Challenger_Expedition (last accessed 6 December 2016).

Cables act as sentinels of the deep ocean by providing information on processes that shape the ocean floor such as landslides and turbidity currents. Indeed, the first direct observation of a turbidity current came from cable breaks caused by the 1929 Grand Banks earthquake.[152] The ever-growing database from cable breaks is guiding new research into the role of turbidity currents in transferring heat, carbon and nutrients to the deep ocean and hence their potential influence on marine ecosystems.[153] In addition, analysis of cable faults world-wide is identifying hazard "hot-spots" such as offshore Taiwan.[154]

Due to their transmission speed, capacity and environmental neutrality, fibre-optic cables and hybrid fibre-optic/power cables form the communications and energy "backbones" of major science observatories as highlighted by the Ocean Networks Canada and the US Ocean Observatories Initiative.[155]

The submarine cable community is also engaging with UN agencies regarding the deployment of sensor-bearing repeaters on commercial telecommunications cables.[156] The concept is to develop systems that continuously monitor ocean temperature, pressure, seabed movement and sound to provide information on ocean change, tsunamis and earthquakes. Discussions are wide ranging, and cover not only technology and science, but also legal and commercial aspects.

Another aspect is the role played by cables in reducing carbon dioxide emissions. A cradle-to-grave study estimated the carbon dioxide (CO_2) budget for a fibre-optic cable from its manufacture, operation, maintenance and recovery for recycling.[157] That information was used to assess CO_2 emissions from a

---

152 Heezen and Ewing (n. 123), 849–873.

153 S.J. Kao, M. Dai, K. Selvaraj, W. Zhai, P. Cai, S.N. Chen, J.Y.T. Yang, J.T. Liu, C.C. Liu and J.P.M. Syvitski, 'Cyclone driven deep-sea injection of freshwater and heat by hyperpycnal flow in the subtropics' (2010) 37 *Geophysical Research Letters* L21702.

154 E. Pope, P.J. Talling and L. Carter, 'Which earthquakes trigger damaging submarine mass movements: insights from a global record of submarine cable breaks?', 2016 *Marine Geology*, at: http://dx.doi.org/10.1016/j.margeo.2016.01.009 (last accessed 6 December 2016).

155 Ocean Networks Canada, 2016, *Discover the Ocean, Understand the Planet, Observations* at: http://www.oceannetworks.ca and Ocean Observatories Initiative, 2017, *The Observatory* at: http://oceanobservatories.org/observatories/ (last accessed 6 December 2016 and 13 May, 2017 respectively).

156 R. Butler, 'Using submarine cables for climate monitoring and disaster warning', *Strategy and Roadmap ITU/WMO/UNESCO IOC Joint Task Force*, 2012, at: http://www.itu.int/en/ITU-T/climatechange/task-force-sc/Pages/default.aspx (last accessed 6 December 2016).

157 C. Donovan, *Twenty thousand leagues under the sea: a life cycle assessment of fibre optic submarine cable systems*. Degree Project SoM EX2009-40 KTH Department of Urban Planning and Environment, Stockholm, 2009, at: www.infra.kth.se/fms.

two-day teleconference between Stockholm and New York, on the one hand, and a face-to-face meeting involving 16,000 km of air travel, on the other. Just 5.7 kg of CO2eq were released by the teleconference compared to 1920 kg of CO2eq from the face-to-face meeting.

6 Submarine Power Cables

Although legally holding the same status, there are important differences between power cables and telecommunication cables.[158] Due to physical depth,[159] weight[160] and length limitations,[161] no power cable has been laid or is planned to be laid in the ABNJ. A proposed HVDC power cable connecting Iceland and the UK is under discussion and, if realized, will be laid along the continental shelf and EEZ of those States and the Faroe Islands. In order to keep the cable depth to less than 1000 m, the proposed cable route presently avoids the ABNJ.[162]

The physical size and footprint of submarine power cables in the ABNJ is unknown, but cables on the geological continental shelf and within the EEZ, typically have diameters of 80 to 150 mm. This contrasts with the much smaller 17–22 mm of deep-ocean telecommunication cables.

With regard to environmental aspects, power cables produce electromagnetic fields (EMF). Until recently, our knowledge of any influence on

---

158 Eccles et al., 'Submarine Power Cables', in *Submarine Cables Handbook* (n. 58).

159 The maximum depth which a power cable has been laid is 1600 metres between Italy and Sicily. By contrast, transoceanic fibre optic telecommunications cables have been laid to depths of 9200 metres. (Squire Patton Boggs ICPC table, *Comparison of Representative Telecommunication Cables and HVDC Cables on the High Seas/ABNJ/BBNJ* (3 March 2016)).

160 The weight of a one metre section of HVDC cable varies between 50 kg (110 pounds) and 60 kg (132 pounds). By comparison, a one metre section of fibre optic telecommunications cable laid in the high seas is about 0.59 kg (1.3 pounds.) The weight of power cables limits the depth to which they can be laid in locations in the area beyond the limits of national jurisdiction because cable ships cannot lay and recover such heavy cables at the depths encountered there. (Squire Patton Boggs ICPC table, *Comparison of Representative Telecommunication Cables and HVDC Cables on the High Seas/ABNJ/BBNJ* (3 March 2016)).

161 The longest HVDC power cable is the 590 km NorNed cable. The longest fibre-optic cables include the 9400 km single segment transpacific, 11,300 km multi segment trans-Pacific and the 25,000 km multi landing Europe-Asia systems. (Squire Patton Boggs ICPC table, *Comparison of Representative Telecommunication Cables and HVDC Cables on the High Seas/ABNJ/BBNJ* (3 March 2016)).

162 H. Hafsteinsdót, 'Presentation on the Iceland-UK HVDC Cable', at workshop *Legal Status of Submarine Cables, Pipelines and ABNJ*, Ankara, Turkey, 7–8 April 2016, Centre for Oceans Law and Policy (COLP), University of Virginia, University of Bergen (Norway), Centre for International Law (CIL) National University of Singapore, and the ICPC; proceedings, including both telecom and power cables, available at: dehukam@ankara.edu.tr.

EMF-sensitive marine biota was limited. This situation reflected a dearth of "real-world" field studies. That situation was compounded by complexities associated with the different responses of various marine organisms to EMFs that also varied in this case with (i) cable voltage, (ii) AC or DC systems, (iii) depth of burial, (iv) cable orientation, and other factors. However, despite these acknowledged complexities, a suite of recent field observations and reviews have come to similar conclusions, namely that for the specific power cables and associated marine organism studied, the EMFs have no clear positive or negative effect on the marine biota and that any change detected reflected the natural variability of the marine communities rather than the influence of a cable.[163] It is also noteworthy that field observations showed that EMFs were tightly constrained about a cable; a condition that is consistent with calculated results thus providing some confidence in the model simulations.

Unease has been expressed about heat loss from submarine power cables and its negative effect on benthic biota. However, it was also noted that there was a lack of field data to form a definite conclusion. Nevertheless, Germany stipulated that for power cables within its EEZ, the maximum allowable

163 E. Andrulewicz, D. Napierska and Z. Otremba, 'The environmental effects of the installation and functioning of the submarine SwePol Link HVDC transmission line: A case study of the Polish Marine Area of the Baltic Sea' (2003) 49 *Journal of Sea Research* 337–345; M.S. Love, M.M. Nishimoto, S. Clark and A.S. Bull, 'Renewable Energy in situ Power Cable Observation', US Department of the Interior, Bureau of Ocean Energy Management, Pacific OCS Region, Camarillo, CA. OCS Study 2016–008, 86 pp, at: http://www.boem.gov/2016-008/, last accessed 6 December 2016; Copping, A., et al., *Annex IV 2016. State of the Science Report: Environmental Effects of Marine Renewable Energy Development Around the World*, at: http://tethys.pnnl.gov/sites/default/files/publications/Annex-IV-2016-State-of-the-Science-Report_LR.pdf, last accessed 6 December 2016; Kuhnz, L. et al., 'Potential impact of the Monterey Accelerated Research System (MARS) cable on the seabed and benthic faunal assemblages', *MARS Biological Survey Report*, 2015, 33pp, with appendices, at: https://www.mbari.org/wp-content/uploads/2016/02/MBARI-Potential-impacts-of-the-Monterey-Accelerated-Research-System-2015.pdf, last accessed 6 December 2016; E. Normandeau, T. Tricas and A. Gill, 'Effects of EMFs from Undersea Power Cables on Elasmobranchs and Other Marine Species', US Department of the Interior, Bureau of Ocean Energy Management, Regulation, and Enforcement, Pacific OCS Region, Camarillo, California, *OCS Study BOEMRE* 2011–09, at: http://tethys.pnnl.gov/publications/effects-emfs-undersea-power-cables-elasmobranchs-and-other-marine-species., last accessed on 6 December 2016; J. Sherwood, S. Chidgey, P. Crockett, D. Gwyther, P. Ho, D. Strong, B. Whitely and A. Williams, 'Installation and operational effects of a HVDC submarine cable in a continental shelf setting: Bass Strait' (2016) 1 *Australia Journal of Ocean Engineering and Science* 337–353, see at: http://www.sciencedirect.com/science/article/pii/S2468013316300316, last accessed on 6 December 2016.

temperature rise at 0.2 m sediment depth, directly above a power cable, was 2K (K=Kelvin). That temperature criterion was regarded enough to maintain the benthos in its natural state.[164] On the basis of field measurements and numerical modelling it has been demonstrated that for the upper 3–5 m of seabed sediments, natural sediment variability and seasonal fluctuations in bottom water temperatures readily exceed the 2K criterion.[165] Heat from power cables was considered to be a second order effect.

Concerns have also been raised about potential pollution associated with oil-insulated submarine power cables.[166] In the case of the ABNJ, the most likely type would be HVDC systems, which since the 1990s, are insulated with either mass impregnated paper or the more widely used cross-linked polyethylene (XPLE). Both of these designs provide better insulation without the environmental risk posed by their oil-bearing predecessors.[167]

IV The Adequacy of Existing International Law and Ocean Governance for Submarine Cables in the ABNJ

General Assembly Resolution 69/292[168] opens up the possibility of an implementing agreement for the area beyond national jurisdiction to address biodiversity issues. A salient aspect of the Resolution 69/292 mandate is that any proposed terms in a possible new implementing agreement must not undermine UNCLOS. In the particular case of international submarine cables this aspect is vital, based on their low impact to the marine environment and historically proven practicality that has provided the world with its amazing critical infrastructure communication system.

The success of the world's submarine cable systems would not have occurred but for the crucial support it has received from UNCLOS (building also upon previous international treaties), which comprehensively addresses rights and obligations related to submarine cables in all of the maritime zones

164 C. Müller, R. Usbeck and F. Miesner, 'Temperatures in shallow marine sediments: Influence of thermal properties, seasonal forcing, and man-made heat sources' (2016) 108 *Applied Thermal Engineering* 20–29.

165 Ibid.

166 OSPAR Commission, 'Background document on potential problems associated with power cables other than those for oil and gas activities', *Biodiversity Series* Publ. No 370/2008.

167 Eccles et al., 'Submarine Power Cables', in *Submarine Cables Handbook* (n. 58), 237–254.

168 UN doc. A/RES/69/292: Development of an international legally-binding instrument under the United Nations Convention on the Law of the Sea on the conservation and sustainable use of marine biological diversity of areas beyond national jurisdiction.

established by UNCLOS. Under UNCLOS, the freedom to lay cables includes those operations associated with this freedom such as cable route surveys and repairs.[169]

Submarine cables in the area beyond the limits of national jurisdiction are seldom disturbed once laid with no burial below the seabed surface. As noted previously, worldwide cable fault records for the period of 2008–2015 show that in the area beyond the limits of national jurisdiction there are on average little more than four faults annually spread out in all of the world's high seas areas. These few faults are generally associated with earthquakes or other natural phenomena since there are few, if any, human generated events at those depths. This data and comparison to faults in territorial seas and EEZs is shown graphically in Annex, below.

Also shown on the graphs are the causes for repair delays for these faults (see Annex, below). There is no requirement for repair permits on the high seas. While permits for emergency repairs to international cables in the EEZ are inconsistent with UNCLOS freedom to lay and maintain cables, several coastal States insist on permits, causing delays with negative results for the resiliency of the world's undersea cable communications. The graphs also underscore the unpredictable patchwork of a minority group of coastal State permitting regimes in their EEZs and the corresponding alarming delay to emergency cable repairs (see Annex, below). These coastal State regulations and actions are classified as excessive maritime claims.[170] Similar excessive permitting would not be helpful to introduce in an instrument on BBNJ to apply in the area beyond the limits of national jurisdiction. This reinforces the wisdom in UNCLOS that allows for the freedom to lay and repair cables on the high seas proper.

A *The Factual Context for International Law for Cables in the ABNJ*

The time tested and proven rules and principles of international law codified in UNCLOS are fundamental to international submarine communications. To understand how well the current UNCLOS provisions work and the risks involved of unintended consequences in altering these balanced provisions, it is important to understand the practical reality about international telecommunication submarine cables.

First, the foremost priority for the submarine cables operation is the integrity and resilience of the submarine cable systems, which are critical for a wide

169 See discussion above in Section II.C.2 and 3.

170 J.A. Roach and R.W. Smith, *Excessive Maritime Claims*, 3rd edn, (Martinus Nijhoff: Leiden/Boston, 2012), at pp. 460–462, citing various States' legislation.

variety of essential services that many take for granted. Additional regulations that undermine traditional freedom to lay and repair cables in areas beyond national jurisdiction would have a detrimental effect on the accessibility and reliability of the cable network.

For context in submarine telecommunication cable reliability and resilience, ocean policy makers, diplomats and scholars should consider seven key points that apply universally:

1. As established in the preceding section of this paper, cables have a neutral environmental footprint on the seabed. In the area beyond the limits of national jurisdiction, submarine cables are surface laid on the flat seabed, not buried; to avoid damage to potential biological "hot spots" they are not laid on the tops or flanks of seamounts and avoid areas of active volcanism. The physical footprint of a fibre-optic cable is tiny, its diameter fitting within the space between these brackets: (   ).
2. There is no single global submarine cable network any more than there is, for example, a single world airline network. (There are about 241 active separate and *decentralized* international cable systems totalling 1,046,138 km of submarine cables).[171]
3. Cable systems are either owned by consortia of four to thirty private companies or in some cases by a single company. About 99% of international telecommunication cables are non-government owned. Cable systems are not "flagged" to any one State.
4. Cable repair arrangements are organized regionally by private contract—not by government mandate. Contracts require repair ships to sail within 24 hours of notice of a cable fault. The goal is to enable fast response and repair.
5. There are about 59 cable ships in the world; about half are on stand-by to carry out emergency repairs pursuant to cable ship pooling contracts with various cable owners and cable ship operators, and the other half are laying new cables or performing other tasks (training, vessel maintenance, out-of-service cable recovery).
6. Cable ships are expensive, custom built, conspicuous, require specialized crews, and fly diverse flags (UK, France, Marshall Islands, Singapore, Japan, China, Korea, UAE, Panama, Denmark, Norway, Spain, Italy, Philippines, Mauritius, Barbados, Belize, Indonesia). This results in competitive rates and efficiency.

171 WFN *SubTel Forum* data base analysis reported to the author (D. Burnett) in an email dated 4 January 2017.

7. Cable repairs are urgent not only to restore service, but because each cable acts as the backup for other cables that are damaged and awaiting repair, thus facilitating their resiliency.

In light of the above points, the most pressing concern for the cable community is the possibility that the existing UNCLOS provisions for submarine cables would be changed or overridden by a possible new environmental regulatory regime implemented under the BBNJ process. Given the critical importance of telecommunication cables to society and economies, together with its neutral to minor environmental footprint, it is difficult to see any valid reasons for modification of the current submarine cable provisions in UNCLOS associated with a new BBNJ implementing agreement.

B *Balancing Submarine Cables and Other Uses with BBNJ*

1 Merchant and Fishing Vessels and Pipeline Crossings

The freedom to lay and maintain cables on the high seas is indeed not unqualified. Always present are the obligations to avoid conduct that prejudices the repair of other cables or pipelines,[172] to indemnify damage to any first laid cable or pipeline that is crossed ("the first laid rule"),[173] to indemnify mariners or vessel owners who, through no fault of their own foul a cable, but sacrifice their gear to avoid injury to the cable,[174] and to show "due regard" for the interests of other States in the exercise of the freedom of the high seas and with respect to activities in the Area.[175]

One NGO suggests a need for strategic environmental assessments to resolve conflicts between fishers and cable-layers on the high seas;[176] but no evidence was cited. First, there are no such conflicts on the high seas as the cable is laid at depths that far exceed commercial fishing. Second, the Cable Convention and COLREGS (in Rule 18) have long established a satisfactory safety distance of one nautical mile between fishing vessels (or any vessel) and cable-layers that display the appropriate day shape or lights while engaged in operations.[177]

---

172 Article 112(2), UNCLOS.

173 Article 115, UNCLOS.

174 Article 114, UNCLOS.

175 Article 87(2), UNCLOS.

176 WWF (Worldwide Fund for Nature), 'Taking a Risk Management Approach to Strategic Environmental Assessments and Environmental Impact Assessments for BBNJ', WWF submission to 2nd PrepCom, August 2016, at: http://www.un.org/depts/los/biodiversity/prepcom_files/WWF_BBNJ_Prep_Com2_2016.pdf, last accessed on 29 November 2016.

177 See above, in Section II.A.

Finally, the cable community and fishers have developed a tried and true working relationships based on international law, and custom and practice that require no new regulation.

Unlike the high seas, in national waters, the most significant largest cause of faults—about 72 to 86%—comes from bottom trawl and similar aggressive fishing activities, as well as from contact with ship anchors.[178] Based on the law-of-the-sea structure as reflected in UNCLOS, the submarine cable community has developed over one and a half centuries sound practices with the fishing and shipping communities including charting, education, liaison, and other time tested techniques that allow these risks to be managed and reduced.[179] In those cases where appropriate, domestic legislation and legal remedies in national admiralty courts[180] are adequate to provide a deterrent to culpably negligent or wilful conduct (excluding terrorism)[181] that threatens or damages the critical international submarine cable infrastructure.

It is emphasized that damage to a telecommunications cable, which is a chemically inert object, results in zero marine pollution, only a disruption in communications.[182] With respect to crossing other cables or pipelines anywhere, the custom and practice of the industries involved allows for these events to take place routinely, safely, and in almost all cases without conflict. The engineers for the crossing systems meet and work out a crossing arrangement or in some cases a formal agreement that complies with the "due regard"

178 R. Wargo and T. Davenport, 'Protecting Submarine Cables from Competing Uses', in *Submarine Cables Handbook* (n. 58), at p. 256.

179 Ibid., 255–279; and ICPC Recommendation No. 6 [Actions for Effective Cable Protection (Post Installation)], Issue 9 (4 November 2015), available upon request at www.iscpc.org/.

180 *Agincourt Steamship Company Ltd v Eastern Extension Australia and China Telegraphy Company Ltd*, 2 KB 305 (1907)(United Kingdom); *Alex Pleven* (France), Whiteman, *Digest of International Law*, Vol. 9, at 948–951; *American Tel & Tel Co v M/V Cape Fear* 763 F Supp. 97 (DNJ 1991) (United States); *Peracomo et al. v Sociéte Telus Communications, Hydro Québec, Bell Canada v Royal and Sun Alliance Insurance Company of Canada*, 2012 FCA 199 (29 June 2012), *aff'd* 2011 FC 494 (2011) (Canada); *The Government of the Netherlands, Post Office v GT Manneteje-Van Dam [Fishing Cutter GO 4]*, File No 325/78 (District Court Rotterdam, decision rendered 20 November 1978), *aff'd sub nom G't Mannethe-Post Office*, File No 69 R/81 and File No rb 325/78 (The Court at the Hague Second Chamber, decision rendered 15 April 1983)(the Netherlands).

181 R. Beckman, 'Protecting Submarine Cables from Intentional Damage–The Security Gap', in *Submarine Cables Handbook* (n. 58), 281–300.

182 *Virginia Commentary* (n. 51), Vol. II, at p. 914, n. 8; and Article 79(3), UNCLOS that allows a coastal State to delineate a pipeline route, but not a cable route.

obligations in UNCLOS.[183] As with fishing and shipping, the current practices and protections for cable and pipeline crossings provided in UNCLOS are adequate and need no additional supplemental treaty provisions or super regulator. Nor is there a need with respect to cables for marine spatial planning since conflicts are historically well managed by those involved.

Instead of a new submarine cable treaty provisions or centralized marine spatial planning by a new or existing entity, greater compliance by States with their existing, but largely unfulfilled, obligations under Article 113 of UNCLOS would strengthen and enhance the reliability of the world's critical ocean infrastructure. The 2015 UN General Assembly resolution on 'Oceans and the Law of the Sea'[184] underscores this point:

> *Recognizing* that fibre-optic submarine cables transmit most of the world's data and communications and hence are vitally important to the global economy and the national security of all States, conscious that these cables are susceptible to intentional and accidental damage from shipping and other activities and that the maintenance, including the repair, of these cables is important, noting that these matters have been brought to the attention of States at various workshops and seminars, and conscious of the need for States to adopt national laws and regulations to protect submarine cables and render their wilful damage or damage by culpable negligence punishable offences, /…/ [*the General Assembly*]
>
> 158. *Also calls upon* States to take measures to protect fibre-optic submarine cables and to fully address issues relating to these cables, in accordance with international law, as reflected in the Convention;
>
> 159. *Encourages* greater dialogue and cooperation among States and the relevant regional and global organizations through workshops and seminars on the protection and maintenance of fibre-optic submarine cables to promote the security of such critical communications infrastructure;
>
> 160. *Also encourages* the adoption by States of laws and regulations addressing the breaking or injury of submarine cables or pipelines beneath the high seas done wilfully or through culpable negligence by a ship flying its flag or by a person subject to its jurisdiction, in accordance with international law, as reflected in the Convention;

183 ICPC Recommendation No. 9A *Telecommunication Cable and Oil Pipeline/Power Cable Crossing Criteria,* Available on request www.iscpc.org.

184 UN doc. A/RES/70/235 (23 December 2015).

> 161. *Affirms* the importance of the maintenance, including the repair, of submarine cables, undertaken in conformity with international law, as reflected in the Convention.[185]

The resolution does not encourage or call upon States to change the provisions in UNCLOS that are involved with submarine cables, but simply to comply with the existing ones.

2 Deep Seabed Mining

The freedom to lay and maintain cables is further qualified by the obligation of all States to exercise this freedom "with due regard for the rights under this Convention with respect to activities in the Area."[186] In this regard, the International Seabed Authority (ISA) has issued a technical study and the ICPC a Recommendation that directly confirms the common obligation of cable owners, the mining contractors, and the ISA to provide notice and meaningful consultation among themselves before initiating their activities.[187] Since 2010, the ISA and the ICPC[188] have productively worked together under a Memorandum of Understanding (MOU) to address a practical "due regard" process. The ICPC hosted a side event on submarine cables at the 22nd ISA Annual Session.

While no mining exploitation licenses have been issued by the ISA, two concession areas have been identified with active submarine cable systems present.[189] Based on the joint workshop approach developed by the ISA and the ICPC, there is a high degree of confidence that any conflicts with these systems or future mining operations and other cables will be professionally resolved by the participants using the existing applicable provisions of UNCLOS and the custom and practice of the submarine cable community in similar natural resource concession crossing situations found in national waters. Accordingly, since the current provisions in UNCLOS are adequate, there is no need for any new implementing treaty to address issues or spatial planning about deep seabed mining and submarine cables in the area beyond the limits of national jurisdiction.

---

185 Ibid., preamble (at 4/53–5/53) and paras. 158–161.

186 Article 87(2), UNCLOS.

187 *Submarine Cables and Deep Seabed Mining: Advancing Common Interests and Addressing UNCLOS "Due Regard" Obligations*, ISA Technical Study: No. 14 (International Seabed Authority: Kingston, 2015), at p. 25 (para. 4); ICPC Recommendation No. 17 *Submarine Cable Operations in Deep Seabed Mining Concessions Designated by the International Seabed Authority*, available on request www.iscpc.org.

188 On the ICPC, see footnote 14, above. The ICPC has a membership drawn from the submarine cable community.

189 *Submarine Cables and Deep Seabed Mining* (n. 187), at p. 10 (para. 4), and Annexes B and C.

3 Reuse of Out-of-Service Submarine Cables

Some NGO's have raised the issue of out-of-service submarine cables as a basis for regulation of international cables in the area beyond the limits of national jurisdiction. In a simplistic but impractical view, the argument is that every cable must be removed at the end of its current 25 to 40 (with upgrade) year design life. Again this is a perceived problem that does not exist as a legal or a practical matter. First, as mentioned, the environmentally neutral to very minor footprint of a garden hose-sized fibre-optic cable on the vastness of the seabed area beyond the limits of national jurisdiction is poor justification to create a regulatory framework or high seas regulator. Adequate legal structures already exists in UNCLOS for rational, realistic and informed decisions on the complex issue of submarine cable recovery.[190]

Second, in an ICPC recommendation "Management of Redundant and Out-of-Service Cables",[191] this matter is already adequately addressed. This recommendation, the product of extensive study and now widespread experience, lays out ten considerations when a submarine cable is taken out-of-service. These include: (1) salvage considerations; (2) safety considerations because of proximity of other objects, cable or pipeline systems, or sensitive areas; (3) present and future environmental impacts on the marine environment; (4) risks of cable shift over time; (5) costs and technical feasibility; (6) deep storage for possible new use for science, artificial reefs, or re-use to connect developing States in a cost effective manner;[192] (7) comparative environmental impact of removal compared to allowing the cable to remain *in situ*; (8) cable protection obligations (sacrificed gear/anchor indemnity); (9) potential socio-economic benefits of cable recovery and (10) deep-sea bed mining notifications. As the above criteria demonstrate, removing a cable is actually a complex decision with many constructive options that requires careful evalu-

---

190 Burnett, 'Out-of-Service Submarine Cables', in *Submarine Cables Handbook* (n. 58), at pp. 213–222.

191 ICPC Recommendation No. 1, Issue 13 (12 January 2016): 'Recovery of Out of Service Cables', available upon request to the ICPC: https://www.iscpc.org/publications/recommendations/.

192 Burnett, 'Out-of-Service Submarine Cables', in *Submarine Cables Handbook* (n. 58), at pp. 214–215, describes successful examples of out-of-service cables donated to scientific institutions for research, used for artificial reefs, or recycled to provide low cost cables to developing States, especially islands whose small populations would likely not justify the expense of a new cable system, and but whose population would immensely benefit from the capacity on the recycled cable system that easily meets local needs. The Gemini-Bermuda cable system is an example of the latter; see Ibid., at p. 214.

ation and study by subject matter experts on a case-by-case basis. There is no simple solution. In some cases, the best result is to leave a cable, or parts of a cable, *in situ*. This is because after decades on the seabed, the cable has basically equilibrated with the marine environment and pulling it up might create unwanted environmental disturbance. In other cases partial recovery for re-use or salvage is the better result. In cases of salvage, there are two ICPC member companies that have the specialized equipment and expertise to safely recover cables in those circumstances, and they have recovered and recycled thousands of kilometres of out-of-service submarine cables.[193] In the case of reuse of submarine cables there are companies that are well versed in these types of operations, which have proven themselves of value, especially to small island developing States.[194]

C *Legal Environmental Aspects of Submarine Cables with* BBNJ

The current UNCLOS balance between submarine cables and the environment and sustainability is easily jeopardized by a potential new regulatory regime characterized by some as an enhanced "conditional freedom of the seas."[195] But enhancing the freedom to lay cables with new conditions impedes the nimble, efficient, and innovative character of the submarine cable community in a spiralling tangle of what appears to be an unnecessary, ever changing, unpredictable and excessive BBNJ regulations imposed by well-intentioned new or existing regulatory entities.

The high risk of unintended consequences of such actions needs to be carefully considered based on the unique known situation of submarine cables in the marine environment. For example, even the indirect centralization and well-intentioned control of the world's submarine cable systems by a BBNJ Environmental Impact Assessment process and similar permission requirements by a new or existing governance entity raises troubling and unpredictable consequences that may well diminish the ability to counter

193 Alcatel Submarine Networks Ltd. (France) and Mertech Marine Pty Ltd. (South Africa).

194 TE Subcom (United States), Global Marine Systems Ltd. (United Kingdom), and Alcatel Submarine Networks Ltd. (France).

195 D. Freestone, 'Modern principles of high seas governance: The legal underpinnings' (2009) 39 *Environmental Policy and Law* 44–49, as cited in D.E.J. Currie and M. Davis, 'Governance Principles Relevant to Marine Biodiversity in Areas Beyond National Jurisdiction: Submission to the Chair' (Submission by the Greenpeace International delegation, 25 February 2016), at 9; available at: http://www.un.org/depts/los/biodiversity/prepcom_files/greenpeace.pdf.

cybersecurity threats to the world's decentralized undersea communication systems, making them more vulnerable to such threats.

Presently each submarine cable system maintains its route position data confidentially, carefully screening requests for access, in most cases limiting disclosure to charting authorities out to depths of 2000 metres as part of cable protection. Potential intruders would have to breach about 241 separate systems to gain access to all of the confidential information for all of these systems. Doing this so many times for each system increases the possibility of timely detection and effective cyber countermeasures. But if a single new or existing UN agency has control of this data, only a single breach is required to access *all* of the sensitive information.

If a new or existing international entity is tasked with centralized coordinating marine spatial planning for everything including cables, it will likely obtain all of the data on the existing submarine cable positions and centralize them in a single database. Terrorists or anarchists, or any other intruders, could then hack into a single target database, dramatically increasing the risk that current cable systems will be compromised in some form.

A new BBNJ implementing regulatory approach for centralized governance—to include that of submarine cables—is not necessary for several reasons. Submarine cables are already well regulated by coastal States through national legislation whenever an international submarine cable lands in the territory of a State or transits its territorial sea. The world's small fleet of cable ships is regulated by their flag States and subject to port and coastal State jurisdiction as well. However, while non-flag State regulatory regimes do not apply on the high seas proper, it is the historical and scientific environmental record for modern submarine cables that speaks for itself. There is no justification in grafting either a new regulatory and governance regime on a proven ongoing process under UNCLOS or the creation of a super-international regulator for submarine cables when no significant environmental harm exists.

The physical footprint of a submarine cable in the area beyond the limits of national jurisdiction is only 17–22 mm in diameter. Using a fibre-optic cable diameter of 22 mm, assuming total fibre-optic cable length for 150 systems in service in the area beyond the limits of national jurisdiction totals approximately 314,350 km,[196] the total ocean coverage by cables is estimated to be about 6.9 km^2. Using an estimated area beyond the limits of national jurisdiction

196 See Table 3 in Section IV.C.2, below: 'Analysis of International Fibre-Optic Submarine Cables in ABNJ MPAs'.

of about 230,000,000 km^2,[197] the percent of that area covered by in-service cables is about 0.00002%.

The amount of power in a submarine telecom cable is a constant DC current of about 0.6 to 1.0 amperes. By comparison, a mid-range laptop computer operates on about 3 amperes and most household circuit breakers are around 10–20 amperes.[198]

In terms of numbers of international cables, the cumulative impact is also minimal. The numbers of these dispersed garden hose-like structures in the vastness of the oceans are regionally small, stable, or experience only small increases in numbers to respond to increased capacity demand. One reason for the number stability, besides the considerable cost of a new cable system, is the new, low cost upgrades that became available around 2006. These upgrades based on further division of the light spectrum, allow the capacity of an existing operational cable to be upgraded by large multiples by simply changing equipment at the cable landing stations, leaving the physical cable on the seabed undisturbed.[199]

Submarine cables are a lawful use of the sea and have now been in the world's oceans from the telegraph era 1850–1950, to the telephone era (1950–1986) to the optical era (1987 to the present).[200] In this span of history, submarine cables have never been associated with an irreversible loss of any species.[201] In fact, out-of-service cables are used as artificial reefs and reused for monitoring of the ocean environment.[202]

---

197 See Table 2 in Section III.A, above.

198 D. Burnett, D. Freestone and T. Davenport, *Submarine Cables in the Sargasso Sea: Legal and Environmental Issues in Areas beyond National Jurisdiction*, report of a workshop held at the George Washington Law School, Washington, D.C., on 23 October 2014, Workshop Report (16 January 2015), para. 49, at p. 20; available at: https://www.iscpc.org/documents/?id=1792.

199 B. Clesca, H. Fevrier, and J. Schwartz, 'Upgrading cable systems? More possibilities that you originally think of!' (2012) 66 *Submarine Telecoms Forum* 23–27.

200 S. Ash, 'The Development of Submarine Cables', in *Submarine Cables Handbook* (n. 58), 19–39; and Appendix 1, 'Timeline of the Submarine Cable Industry', in Ibid., at 377–392.

201 H.D. Smith, J.L. Suárez de Vivero and T.S. Agardy, *Routledge Handbook of Ocean Resources and Management* (Routledge: London/New York, 2015), at p. 353.

202 Burnett, 'Out-of-Service Submarine Cables', in *Submarine Cables Handbook* (n. 58), at pp. 214–215.

The 2015 Oceans and the Law of the Sea Report of the Secretary-General,[203] in its paragraph 54 succinctly sums up the conventional wisdom about international submarine cables and the marine environment:

> The environmental dimension of submarine cables is, however, less apparent. Submarine cables themselves are considered to have a low-carbon footprint and a small relative impact on the environment, with the maintenance of submarine cables causing the highest impacts as a result of the operation of the cable ships themselves.[87][204] Submarine cables have the potential to be [*sic*] contribute actively to disaster warning and addressing climate change, with work under way to examine the potential for monitoring purposes.[88][205]

The small effect of submarine cables in the environment is documented in a long record of collaborative study in well established, peer review international scientific journals, workshops, and studies with well regarded scientists and legal scholars.

In 2009 the United Nations Environment Programme (UNEP), the World Conservation Monitoring Centre (WCMC) and the ICPC collaborated and published a cross-disciplinary review and study of the impact of submarine cables in the marine environment. The 64 page UNEP-WCMC Report, based on a review of 191 cited peer reviewed scientific, academic, industry and government studies, and vetted by 18 external reviewers, concluded: "as outlined in this report, the weight of evidence shows the environmental impact of fibre-optic cables is neutral to minor."[206] Since the UNEP-WCMC milestone report, approximately 26 other peer-review studies have been completed on various environmental aspects including natural hazards, cable stability, seabed recovery from cable burial, marine mammal and shark interactions, EMF and heat outputs. The cumulative result of these studies (many of which are listed in the footnotes

---

203 UN doc. A/70/74 (30 March 2015).

204 Ibid., footnote 87 in the report, which reads: 'C. Donovan, "Twenty thousand leagues under the sea: a life cycle assessment of fibre optic submarine cable sytems", (Stockholm, 2009)'.

205 Ibid., footnote 88 in the report, which reads: 'See A/67/79/Add.1 and A/69/71/Add. 1. See also www.itu.int/en/ITU-T/climatechange/task-force-sc/Pages/default.aspx.'

206 Carter et al. (n. 64), at p. 54.

to this monograph) echoes the UNEP-WCMC report that modern submarine cables have a neutral to minor impact in the marine environment.[207]

A subsequent 2015 ISA publication, *Submarine Cables and Deep Seabed Mining,* notes that "submarine cables have a reduced carbon footprint" and that "their environmental impact is minor if not negligible."[208]

In 2015 an interdisciplinary workshop "Submarine Cables in the Sargasso Sea: Legal and Environmental Issues in Areas beyond National Jurisdiction"[209] there was a convergence of views on several relevant ABNJ findings:

a. For water depths over 2000 m, cables are laid directly on the seabed and hence seabed disturbance is minimal.
b. The laying of a cable is intended to be a one-off operation in the 25 year design life of a cable, although faults may occur mainly via natural and human-related hazards. When repairs are needed, grapnels used for cable recovery may disturb the seabed along meter-wide paths. The recovered cable is repaired and lowered to the seabed to minimize further disturbance. Again, a repair is planned to be a one-off operation in a cables' remaining design life.
c. Cable operations also have a low carbon footprint, and are undertaken with concern for safety, fuel economy and the environment.
d. Once the cable is laid, the physical impact on the seabed is minimal. The size of communications cables is small, ranging from 17 mm to 22 mm. Cables are protected by a substantial sheath of marine grade polyethylene, which is inert in the ocean.
e. Research into cables and benthic organisms living on and in the seabed show that there is no statistical difference in the abundance and diversity for organisms living near and away from a cable.

207 See *Submarine Cables Handbook* (n. 58), especially Chapter 7 (L. Carter, D. Burnett and T. Davenport, 'The Relationship between Submarine Cables and the Marine Environment', at p. 202), Chapter 10 (Carter, 'Submarine Cables and Natural Hazards') and Chapter 14 (L. Carter and A.H.A. Soons, 'Marine Scientific Research Cables', 323–349). In this treatise, the low environmental footprint of submarine cable was reconfirmed where it is observed that in practice cables are laid to avert environmental harm by identifying during the cable route survey process fragile ecosystems that are bypassed.

208 *Submarine Cables and Deep Seabed Mining* (n. 187), at p. 18 and Annex F (Environmental Impacts of Submarine Cables).

209 Burnett et al., *Submarine Cables in the Sargasso Sea* (n. 198).

f. Studies have also been done on the direct environmental impact of cables on marine life[210] (including whales[211] and sharks[212]). Whale entanglements with cables ceased with the transition from telegraph to coaxial cables by the early 1960s, which reflected improved cable design, laying techniques, burial, and seabed mapping.

While the Sargasso Sea workshop focused on the Atlantic Ocean, the above findings are consistent with the industry custom and practice in areas beyond the limits of national jurisdiction in other oceans. These findings, as well as the neutral nature of submarine cable impact to the marine environment, were again confirmed in 2016 at an international workshop in Ankara, Turkey.[213]

The United Nations World Ocean Assessment reviewed submarine telecommunications cables and concluded that they "have very limited environment impacts."[214] It also acknowledges the socio-economic importance of cables and the role played by the ICPC in ensuring the safety of cables and maintaining their nil to minor effect on the environment.

Notwithstanding the thorough and consistent record of scientific and academic review of modern cables discussed above, a remark about the precautionary approach is appropriate. The unique status and cumulative studies of submarine cables make, however, the application of the precautionary

210 See, e.g., I. Kogan, C. Paull, L. Kuhnz, E. Burton, S. Von Thun, H.G. Greene and J. Barry, 'ATOC/Pioneer Seamount cable after 8 years on the seafloor: Observations, environmental impact' (2006) 26 *Continental Shelf Research* 771–787; and B.M. Grannis, *Impacts of mobile fishing gear and a buried fibre-optic cable on soft-sediment benthic community structure*, MSc thesis, University of Maine, 2001, 100 pp), as well as E. Andrulewicz, D. Napierska and Z. Otremba, 'The environmental effects of the installation and functioning of the submarine SwePol Link HVDC transmission line: A case study of the Polish Marine Area of the Baltic Sea' (2003) 49 *Journal of Sea Research* 337–345, that all show no negative impact of submarine power cables on the abundance and diversity of benthic organisms.

211 M.P. Wood and L. Carter, 'Whale Entanglements with Submarine Telecommunications Cables' (2008) 33 *IEEE Journal of Oceanic Engineering* 445–450.

212 L.J. Marra, 'Shark bite on the SL Submarine Light Wave Cable System: History, Causes, and Resolution' (1989) 14 *IEEE Journal of Oceanic Engineering* 230–237.

213 'Legal Status of Submarine Cables, Pipelines and ABNJ', Ankara, Turkey 7–8 April 2016, sponsored by Research Center of the Sea and Maritime Law (Dehukam), Ankara University, Centre for Oceans Law and Policy (COLP), University of Virginia, University of Bergen, Norway, Centre for International Law (CIL) National University of Singapore, and the ICPC, planned publication of proceedings, including both telecom and power cables, is available at: dehukam@ankara.edu.tr.

214 United Nations World Ocean Assessment, see at: http://www.worldoceanassesment.org.

approach as a basis for new regulation of submarine cables in the BBNJ process inappropriate.

The noted scientist Professor Sir Peter Gluckman, chief Science Advisor to the New Zealand government, observed:

> The Precautionary Principle was initially intended as a framework FOR ACTION in the face of scientific uncertainty—that is, not using the absence of evidence as reason not to act—for example on climate change. But Callon pointed out that, when applied to the innovation space, the Precautionary Principle was being wrongly framed as a reason for abstention and inaction. /…/ The default position has insidiously shifted to an interpretation that allows nothing in the face of any uncertainty, which by definition must exist. And so the misuse of this principle has become a guiding tool for advocates trying to stop any particular innovation.[215]

Any consideration of changes to UNCLOS must first carefully consider that innovation allowed by the existing UNCLOS provisions on the freedom to lay and maintain international cables is the core lifeblood of the modern internet and international communications. The *cloud* is in fact thriving because of the submarine cables that link data centre servers and peoples in all nations. The existing environmental requirements in UNCLOS are sufficient to safeguard the marine environment from the demonstrably nil to very small environmental risks posed by the small, chemically inert submarine cables.

The precautionary approach should not be mechanically applied to justify regulation of well-established routine international cable operations and routes. As documented in this paper and the abundant literature cited, cable operations are known activities with a long historical record of safe interaction with the marine environment. These routine operations have no significant impact on the marine environment and certainly not one that would justify application of the precautionary approach to them. A new and untried regulatory regime may stifle and suffocate the innovation that has given the world its critical international cable infrastructure. This could lead to unintended consequences such as limiting or prolonging the efforts to bring fibre-optic cables to small islands and developing States or providing alternate cable routes and restoration options to minimize risks from natural disasters. Such

215 P. Gluckman, 'The place of science in environmental policy and law', The Salmon Lecture to the Resource Management Law Association, Wellington, New Zealand, 2 September 2015, at pp. 5–6, available at: http://www.pmcsa.org.nz/wp-content/uploads/Salmon-Lecture_Final.pdf.

consequences, delays in repairs, delays in connecting small island nations as well as African nations, and providing alternate cable redundancy for security have already been the reported focus of international leaders at high levels of the Organisation for Economic Co-operation and Development (OECD)[216] and the International Telecommunications Union (ITU).[217]

1 Environmental Impact Assessments for Submarine Cables

Under UNCLOS, submarine cables in the area beyond the limits of national jurisdiction are already subject to provisions within Part XII on "Protection and Preservation of the Marine Environment." In the context of its Section 4, on "Monitoring and Environmental Assessment", Article 206 [Assessment of potential effects of activities] provides that when States that have reasonable grounds for believing that planned activities under their jurisdiction or control may cause substantial pollution of or significant and harmful changes to the marine environment they shall, as far as practicable assess the potential effects of such activities on the marine environment. Thus the flag State of the cable ship or a State whose nationals own or operate an international submarine cable already have authority needed to carry out an environmental impact assessment (EIA) if justified.

By definition submarine cables are not "pollution of the marine environment," nor can cables realistically cause such pollution. A modern fibre-optic cable is not a substance or energy likely to result in deleterious effects as harm to living resources and marine life.[218] As demonstrated previously, submarine cables also do not cause "significant harmful changes"[219] to the marine environment. In view of the substantial scientific record, Article 206 has not been applied on the high seas. The point is, however, it remains a legal obligation that is in force and always available to States to protect the marine environment.

216 Broadband and the Economy, Ministerial Background Report, DSTI/ICCP/IE(2007)3/FINAL, 17–18 June 2008, at pp. 12–13 ('What happens when the cables fail') and at pp. 32–33 ('The physical infrastructure of broadband and ICT-enabled trade in services').

217 *Broadband: A Platform for Progress. A report by the Broadband Commission for Digital Development*, ITU/UNESCO, June 2011, at p. 44 ('Although Africa has, at present, low penetration of broadband, new initiatives are taking place to improve connectivity. These include national plans for background infrastructure and new links via submarine cables'); and at pp. 155 ('A key indicator for broadband development in Africa is the deployment of basic infrastructure, such as international fibre-optic cables. Many African governments have co-sponsored new cables of this type along the continent's east and west coasts, with the aim of improving broadband connectivity'.).

218 Article 1(4), UNCLOS.

219 Article 206, UNCLOS.

In view of Article 206, there is no need to create a new overlapping EIA obligation in a BBNJ implementing agreement for submarine cable laying and repair. As the UNEP-WCMC report highlights:

> EIAs for cable operations are rare and are generally limited to a coastal State's territorial sea. The European Union EIA Directive currently does not explicitly impose an EIA requirement on cable-laying projects.[220]

The Sargasso Sea workshop in particular extensively looked at all aspects of cables on the high seas and their relationship to the marine environment and the convergence of views reached was "that the impact of submarine cables and cable operations in the deep water of the Sargasso Sea would also be minimal" because cables were not buried or laid on seamounts, hazards posed by turbidity currents were slight and hence repairs were infrequent.[221] Furthermore, participants in the workshop agreed that there was no baseline in the deep ocean by which to compare any change in the marine environment and no cost benefit analysis to justify the delays and costs associated with an EIA in ABNJ where there was no "clear benefit of EIAs in such areas".[222]

"It was also agreed that an EIA (or equivalent) should not be required before cable repairs take place in the Sargasso Sea given the importance of ensuring that repairs are done as expeditiously as possible."[223] Where coastal States have required an EIA for submarine cables, the normal time frame varies from weeks to years to carry out the EIA and submit it to the government authority making the request.[224] In the case of an emergency cable repair, this would be like having the fire department carry out an EIA on a burning building before attempting to bring the fire under control and extinguish it. For laying a new cable system, new delays for EIA in high seas or in the seabed beyond national jurisdiction would threaten the viability of a project that depends on being innovative and nimble to increasing broadband demand and at the same time compliant with a budget, financing, and project timeline. In view of the demonstrated lack of any significant harm from these activities, there is no need for such new regulatory and untested burden on submarine cables.

---

220 Carter et al. (n. 64), at p. 30.

221 Burnett et al., *Submarine Cables in the Sargasso Sea* (n. 198), at p. 20.

222 Ibid., at p. 22.

223 Ibid., at p. 23.

224 Carter et al., 'The Relationship between Submarine Cables and the Marine Environment', in *Submarine Cables Handbook* (n. 58), at pp. 200–201.

We consider Article 206 to be sufficient to safeguard the marine environment in the case of submarine cables in the area beyond the limits of national jurisdiction. As such, we see no reasonable justification to add to submarine cable operations, especially emergency repairs, a new obligation to carry out an EIA under any new undefined BBNJ regulatory regime.

Some NGOs argue that there can be no exceptions for any industry to new EIA mandates in a possible BBNJ implementing convention, and that a new international organization be created to oversee and approve EIAs for all ocean activities.[225] A basic flaw in this somewhat dogmatic approach is that different activities do not all fit into the same rigid legal containment mechanism, nor is it reasonable to assume one regulator would possess the necessary expertise to oversee so many diverse activities. Submarine cables, deep seabed mining, oil and gas exploration, and fishing are all activities carried out by different industries in the oceans. But each industry has its own technology, environmental impacts, rules, and practices that are essential to it. Some industries may require a specialized arbitrator or a judge to ensure fair play. (i.e., deep seabed mining with contractors and the ISA and their relationship to the Seabed Disputes Chamber of International Law of the Sea Tribunal).[226] Painting all activities with the same regulatory brush administered by a single international regulator will engender the bureaucratic regulatory overkill that will stifle innovation and likely also lack practical utility. Each activity, especially submarine cables, should be judged on its environmental record and any new treaty EIA requirement should recognize these different outcomes.

2 Marine Protected Areas in the ABNJ

Historically, submarine cables have co-existed in MPAs with no significant harm to the environment.[227] In fact, scientists have concluded that cable protection zones with the appropriate environmental attributes such as rocky reefs to encourage fish aggregation, can make ideal *de facto* marine protected areas.[228]

---

225 *Earth Negotiation Bulletin*, International Institute of Sustainable Development, Vol. 25, No. 103, PrepCom 1 # 8, 6 April 2016, at: http://www.iisd.ca/download/pdf/enb25103e.pdf (last accessed 11 January 2017).

226 Annex VI, Article 14, UNCLOS.

227 Carter et al., 'The Relationship between Submarine Cables and the Marine Environment', in *Submarine Cables Handbook* (n. 58), at pp. 202–207; Smith et al., *Routledge Handbook* (n. 201), at p. 360.

228 Carter et al., 'The Relationship between Submarine Cables and the Marine Environment', in *Submarine Cables Handbook* (n. 58), at p. 207; Carter et al. (n. 64), at p. 37.

The majority of submarine cable routes follow the same "tried and true" historic paths of earlier telegraph cables.[229] This reflects the proven low environmental impact and natural hazard risks historically experienced with submarine cable laying along these routes. As a result new MPAs may well be considered over existing cable routes. Submarine cables and MPAs are not mutually exclusive by any means.

The following table underscores that submarine cables are mutually consistent with MPAs and have been so for years.

TABLE 3 *Analysis of international fibre-optic submarine cables*[230] in *ABNJ MPAs*[231]

Data description	ABNJ area
Total number of cable systems in database in ABNJ	150
Total cable systems in MPAs	22
Percent of cables that cross MPAs	15%
Total km of cables in ABNJ in data base	314,350 km
Total km in ABNJ MPAs	5,362 km
Percent of total km in MPAs	1.7%

Five new high seas MPAs[232] have been declared under the Convention for the Protection of the Marine Environment of the North-East Atlantic (OSPAR).[233] Concerns have been raised by the submarine cable community when OSPAR unilaterally and without any consultation with submarine cable stakeholders declared high seas MPAs over five existing transatlantic cable systems.[234]

229 N. Starosielski, *The Undersea Network* (Duke University Press, 2015), at p. 2.

230 Includes data supplied by Global Marine Systems Limited; Copyright [2016] Global Marine Systems Limited. This data or information is provided on a reasonable endeavours basis and Global Marine Systems Limited does not guarantee its accuracy or warrant its fitness for any particular purpose. Such data or information has been reprinted with the permission of Global Marine Systems Limited.

231 MPAtlas *How much of our ocean is protected?* (Marine Conservation Institute: Seattle, WA., 2016), at: http://www.mpatlas.org./. Last accessed 14 May 2017.

232 For general descriptions of the MPAs, see at: http://mpa.ospar.org/home_ospar/mpa_datasheets (last accessed 3 March 2017).

233 Published in 2354 *UNTS* 67; reprinted in 32 *ILM* 1069 (1993).

234 Ibid., at pp. 208–212.

However to date the declaration appears inactive for submarine cables with no impact or restrictions on the laying and repair of these systems.[235]

If a high seas MPA was to become a "no go" or restricted area for international submarine cables or if repairs to existing cables are delayed or prejudiced, then of course it could be expected that the cable owners and/or cable ship operators would request their respective States to take up the issue diplomatically to prevent endangering this global critical infrastructure. These remedies already work in UNCLOS and should not be undermined by a new super-international regulator with remit over international cables.

Conclusion

Based on the analysis undertaken in this study, we consider that the UNCLOS obligation of due regard is sufficient to ensure a harmonious relationship between submarine cables and any high seas MPA that may appear in the future.[236] The careful existing balance in areas beyond the limits of national jurisdiction between the environmental considerations (Articles 192 and 206) and the freedom to lay and maintain international submarine cables (Articles 87, 112–115 and 297) has worked successfully since UNCLOS came into force. For that reason, whatever the outcome of the BBNJ process for MPAs and EIAs, the existing structure of UNCLOS for submarine cables should in our view not be undermined by changes to existing time-tested practices that have served humankind so well or by subjecting cables to a new regulatory BBNJ regime that is not required to protect the marine environment from cables. Put another way, there is no need to fix something that is not broken.

It would therefore be recommendable that the BBNJ process does not change or condition the existing provisions in UNCLOS that deal with submarine cables and does not impose any new and additional EIA and MPA requirements for cables in any new implementing agreement. There is precedent for this approach. The drafters of the 2001 Convention on the Protection of the Underwater Cultural Heritage,[237] after considering the historical record that submarine cables had never been a threat or damaged underwater cultural

235 The ICPC now has observer status with OSPAR and is working towards providing information and knowledge to OSPAR about submarine cables and accurate considerations for best environmental practices.

236 Article 87(2), UNCLOS.

237 41 *ILM* 37 (2002) (entered into force 2 January 2009).

heritage, exempted submarine cables from that treaty.[238] The compelling scientific evidence, the long positive track record of submarine cables in the marine environment, and the vital role of cables as critical international infrastructure merit a similar result in the BBNJ process.

Submarine cables, with their small footprint, positive contribution to reducing greenhouse gases, and well-studied neutral to minor environmental impact, stand uniquely apart from high impact uses that are of concern to the area beyond the limits of national jurisdiction such as shipping, deep seabed mining, fishing, pipelines and energy. The knowledge presented in this monograph is unequivocal in concluding that submarine cables should be expressly excluded from any new BBNJ implementing agreement.

References

Agincourt Steamship Company Ltd. v Eastern Extension, Australia and China Telegraph Company Ltd. [1907] 2 K.B. 305 (United Kingdom).

Alex Pleven (France), Whiteman, *Digest of International Law*, Vol. 9, at 948–951 [FN180 format].

'Alloy Angels', *The Economist*, 28 May 2016.

American Telephone and Telegraph v M.V. Cape Fear, 763 F. Supp. 97 (D.N.J. 1991), revised on other grounds, 967 F.2d. 864 (3rd Cir. 1992) (United States) [FN 84 format].

American Tel & Tel Co v M/V Cape Fear 763 F Supp. 97 (DNJ 1991) (United States) [FN 180 format].

Andrulewicz, E., D. Napierska and Z. Otremba, 'The environmental effects of the installation and functioning of the submarine SwePol Link HVDC transmission line: A case study of the Polish Marine Area of the Baltic Sea' (2003) 49 *Journal of Sea Research* 337–345.

Arbitration between Concert Global Network Services, Ltd., in its own capacity, and as co-maintenance authority of submarine cable system TAT-10, as Claimant and Tyco Telecommunications (U.S.) Inc. as Respondent (Arb. New York, SMA 3779, 2002) (United States).

Argo, 2016. *About Argo*, at: http://www.argo.ucsd.edu/.

Broadband: A Platform for Progress. A report by the Broadband Commission for Digital Development, ITU/UNESCO, June 2011.

Burnett, D., 'The 1884 International Convention for Protection of Submarine Cables Provisions Not in UNCLOS Deserve Attention Now', Workshop on the Protection

238 Burnett, 'Out-of-Service Submarine Cables', in *Submarine Cables Handbook* (n. 58), at p. 200.

of Submarine Cables, Singapore, 14–15 April 2011, sponsored by the Centre for International Law, National University of Singapore and the International Cable Protection Committee, http://cil.nus.edu.sg/powerpoint-presentations-and-papers-presented-at-the-2011-workshop/.

Burnett, D.R., R.C. Beckman and T.M. Davenport (eds), *Submarine Cables: the Handbook of Law and Policy* (Brill/Martinus Nijhoff: Leiden/Boston, 2014).

Burnett, D., D. Freestone and T. Davenport, *Submarine Cables in the Sargasso Sea: Legal and Environmental Issues in Areas beyond National Jurisdiction*, report of a workshop held at the George Washington Law School, Washington, D.C., on 23 October 2014, Workshop Report (16 January 2015), at: https://www.iscpc.org/documents/?id=1792.

Butler, R., 'Using submarine cables for climate monitoring and disaster warning', *Strategy and Roadmap ITU/WMO/UNESCO IOC Joint Task Force*, 2012, at: http://www.itu.int/en/ITU-T/climatechange/task-force-sc/Pages/default.aspx.

Carter, L., D. Burnett, S. Drew, G. Marle, L. Hagadorn, D. Bartlett-McNeil and N. Irvine, *Submarine Cables and the Oceans—Connecting the World*, UNEP–ECMC Biodiversity Series No. 31, 2009.

Carter, L., R. Gavey, P.J. Talling and J.T. Liu, 'Insights into submarine geohazards from breaks in subsea telecommunication cables' (2014) 27(2) *Oceanography* 58–67, at: http://dx.doi.org/10.5670/oceanog.2014.40.

Cattaneo, A., N. Babonneau, G. Ratzov, G. Dan-Unterseh, K. Yelles, R. Bracène, B. Mercier de Lèpinay, A. Boudiaf and J. Déverchère, 'Searching for the seafloor signature of the 21 May 2003 Boumerdès earthquake offshore central Algeria' (2012) 12 *Natural Hazards Earth System Science* 2159–2172.

Challenger Society for Marine Science, 2016. *History of the Challenger Expedition* at: http://www.challenger-society.org.uk/History_of_the_Challenger_Expedition.

Chesnoy, J., 'Back Reflection' (2016) 88 *Subtelcom Forum*.

Churchill, R.R., and A.V. Lowe, *The Law of the Sea*, 3rd edn (Manchester University Press, 1999).

Ciena. 'Peak Mobile Bandwidth Per User To Increase Fivefold By 2018', *Ciena Study*, 2 December 2015, at: http://media.ciena.com/documents/2015_02_12_Ciena_ACG_Research_Mobile_FINAL.pdf.

Clesca, B., H. Fevrier, and J. Schwartz, 'Upgrading cable systems? More possibilities that you originally think of!' (2012) 66 *Submarine Telecoms Forum* 23–27.

Coghlan, A., 'Hacker, the humpback whale who got entangled in an internet cable', (19 November 2016) 3100 *New Scientist*.

Collins, K., *Isle of Man Cable Study—preliminary material environmental impact studies*, Preliminary Report, University of Southampton, 2007 (prepared for BT, Global Marine Systems Ltd and Department of Environment, Food and Agriculture Isle of Man).

Convention on the International Regulations for Preventing Collisions at Sea, 1050 *UNTS* 16 (1972) [COLREGS].

Convention on the Protection of the Underwater Cultural Heritage, 41 *ILM* 37 (2002).

Conventions on the Law of the Sea: Hearing before the Committee on Foreign Relations, United States Senate, 86th Cong., 2nd Sess., 20 January 1960 (US Government Printing Office: Washington DC, 1960).

Cook, L., Principal TMT Advisor, International Finance Corporation Telecoms, Media & Technology Venture Group, cited in *Submarine Telecoms Industry Report*, Issue 4, (November 2015).

Copping, A., et al., *Annex IV 2016. State of the Science Report: Environmental Effects of Marine Renewable Energy Development Around the World*, at: http://tethys.pnnl.gov/sites/default/files/publications/Annex-IV-2016-State-of-the-Science-Report_LR.pdf.

Davenport, T.M., "Submarine Communications Cables and Law of the Sea: Problems in Law and Practice," *Ocean Development and International Law* 43:3, 201–242 (2012).

de Juvigny, A.L., T.M. Davenport, D.R. Burnett and D. Freestone, 'Submarine telecommunication cables in the Sargasso Sea' (2015) 30 *International Journal of Marine and Coastal Law* 371–378.

Dees-Gross, F., World Bank Country Director for Papua New Guinea, Timor Leste and the Pacific Islands, quoted in "Underwater Cable to Bring High Speed Internet to Samoa," World Bank Press Release (19 June 19, 2015), http://www.worldbank.org/en/news/press-release/2015/06/19/underwater-cable-to-bring-high-speed internet-to-samoa.

Dengler, A.T., P. Wilde, E.K. Noda and W.R. Normark, 'Turbidity currents generated by Hurricane IWA' (1984) 4 *Geomarine Letters* 5–11.

DOALOS, 'Frequently Asked Questions: Q7: What regime applies to the cables and pipelines?', at: http://www.un.org/depts/los/LEGISLATIONANDTREATIES/frequently_asked_questions.htm.

Donovan, C., *Twenty thousand leagues under the sea: a life cycle assessment of fibre optic submarine cable systems*. Degree Project SoM EX2009-40 KTH Department of Urban Planning and Environment, Stockholm, 2009, at: www.infra.kth.se/fms.

Dupuy, R.J., and D. Vignes (eds), *A Handbook on the New Law of the Sea*, Vol. 2 (Martinus Nijhoff: Dordrecht /Boston/Lancaster, 1991).

Dynamic Load Monitoring, UK, 2016. *Grapnel Data Sheet*, at: http://www.dlm-uk.com/wp-content/uploads/2014/07/Grapnels-datasheet.pdf.

Earth Negotiation Bulletin, International Institute of Sustainable Development, Vol. 25, No. 103, PrepCom 1 # 8, 6 April 2016, at: http://www.iisd.ca/download/pdf/enb25103e.pdf.

E-mail interview with Tony Whitton (http://rosiefigi.com/en_US) (21 June 2016).

Emu Ltd, *Subsea cable decommissioning: A limited environmental appraisal*. Report no 04/J/01/06/ 0648/0415 (2004). Open file report available from European Subsea Cables Association, at: http://www.escaeu.org/.

Flores v Southern Peru Copper, 343 F.3d 140 (2nd Cir. 2003) (United States).

Gavey, R., L. Carter, J. Liu, P. Talling, R. Hsu, E. Pope and G. Evans, 'Frequent sediment density flows during 2006 to 2015, triggered by competing seismic and weather events: Observations from subsea cable breaks off southern Taiwan' (2016) *Marine Geology*, at: doi:10.1016/j.margeo.2016.06.001.

Geneva Convention on the Continental Shelf, 29 April 1958, 499 *UNTS* 311.

Geneva Convention on the High Seas, 29 April 1958, 450 *UNTS* 11.

Gluckman, P., 'The place of science in environmental policy and law', The Salmon Lecture to the Resource Management Law Association, Wellington, New Zealand, 2 September 2015, at: http://www.pmcsa.org.nz/wp-content/uploads/Salmon-Lecture_Final.pdf.

Grannis, B.M., *Impacts of mobile fishing gear and a buried fibre-optic cable on soft-sediment benthic community structure*. M.Sc. thesis, University of Maine, 2001.

Green, M., and D. Burnett, 'Security of International Cable Infrastructure: Time to Rethink?', *Challenges in Maritime Security*, Center for Oceans Law and Policy (2008).

Hafsteinsdót, H., 'Presentation on the Iceland-UK HVDC Cable', at workshop *Legal Status of Submarine Cables, Pipelines and ABNJ*, Ankara, Turkey, 7–8 April 2016, Centre for Oceans Law and Policy (COLP), University of Virginia, University of Bergen (Norway), Centre for International Law (CIL) National University of Singapore, and the ICPC; at: dehukam@ankara.edu.tr.

Hall, I.R., I.N. McCave, N.J. Shackleton, G.P. Weldon, and S.E. Harris, 'Glacial intensification of deep Pacific inflow and ventilation' (2001) 412 *Nature* 809–812.

Heezen, B.C., 'Whales entangled in deep sea cables' (1957) 4 *Deep-Sea Research* 105–115.

Heezen, B.C., and M. Ewing, 'Turbidity currents and submarine slumps, and the 1929 Grand Banks earthquake' (1952) 250 *American Journal of Science* 849–873.

Hollister, C.D., and I.N. McCave, 'Sedimentation under deep-sea storms' (1984) 309 *Nature* 220–225.

ICPC (International Cable Protection Committee). ICPC Ocean Observation Sites and areas-2009; at www.iscpc.org.

ICPC Recommendation No. 1, Issue 13 (12 January 2016): 'Recovery of Out of Service Cables', available upon request to the ICPC: https://www.iscpc.org/publications/recommendations/.

ICPC Recommendation No. 9A Telecommunication Cable and Oil Pipeline/Power Cable Crossing Criteria, available upon request www.iscpc.org.

ICPC. 'About Power Cables', ICPC website, www.iscpc.org at Publications (2012).

ICPC, "Telecommunication Cable and Oil Pipeline/Power Cable Crossing Criteria," Recommendation No. 3, Issue 10A, available at no charge upon request to the ICPC https://www.iscpc.org/publications/recommendations/.

ICPC Recommendation No. 17 *Submarine Cable Operations in the Deep Seabed Mining Concessions Designated by the International Seabed Authority*, available on request www.iscpc.org.

ILC, Yearbook of the International Law Commission, Volume IIA, UN Doc. A/CN.4/79 (1954) at 152.

IMO, Sub-Committee on Navigation, Communications and Search and Rescue, Document NCSR 3/1/2 (January 14, 2016) https://edocs.omo.org/Final/Documents/English/NCSR 3-1-2(E).docx. IMO Sub-Committee on Navigation, Communications and Search and Rescue document NCSR 3/1/2, 14 January 2016, at: https://edocs.omo.org/Final Documents/English/NCSR 3-1-2.

In the Matter of the Chagos Marine Protection Area Arbitration, before an Arbitral Tribunal Constituted under Annex VII UNCLOS, between the Republic of Mauritius and the United Kingdom of Great Britain and Northern Island, Permanent Court of Arbitration, Award of 18 March 2015.

Intergovernmental Panel on Climate Change, *Climate Change 2013: The Physical Science Basis. Working Group I Contribution to the IPCC 5th Assessment Report - Changes to the Underlying Scientific/Technical Assessment* (IPCC-XXVI/Doc.4), at: http://www.ipcc.ch/report/ar5/wg1/#.Uq5tSxaWfHg.

International Convention for the Protection of Submarine Cables, 14 March 1884, T.S. 380.

Kao, S.J., and J.D. Milliman, 'Water and sediment discharge from small mountainous rivers, Taiwan: The roles of lithology, episodic events, and human activities' (2008) 116 *Journal of Geology* 431–448.

Kao, S.J., M. Dai, K. Selvaraj, W. Zhai, P. Cai, S.N. Chen, J.Y.T. Yang, J.T. Liu, C.C. Liu and J.P.M. Syvitski, 'Cyclone driven deep-sea injection of freshwater and heat by hyperpycnal flow in the subtropics' (2010) 37 *Geophysical Research Letters* L21702.

Kogan, I., C. Paull, L. Kuhnz, E. Burton, S. Von Thun, H.G. Greene and J. Barry, 'ATOC/Pioneer Seamount cable after 8 years on the seafloor: Observations, environmental impact' (2006) 26 *Continental Shelf Research* 771–787.

Kordahi, M.F., and S. Shapiro, 'Worldwide trends in submarine cable systems', *Proceedings SubOptic 2004*, Monaco, 2004, 3 pp, at: http://www.suboptic.org go to Resources/ Conference Archives/SubOptic 2004.

Kordahi, M.E., S. Shapiro, and G. Lucas, 'Trends in submarine cable system faults', *Proceedings SubOptic 2007*, Baltimore, 2007, 4 pp, at: http://www.suboptic.org go to Resources/Conference Archives/SubOptic 2007.

Kuhnz, L. et al., 'Potential impact of the Monterey Accelerated Research System (MARS) cable on the seabed and benthic faunal assemblages', 2015 *MARS Biological Survey Report*, at: https://www.mbari.org/wp-content/uploads/2016/02/MBARI-Potential-impacts-of-the-Monterey-Accelerated-Research-System-2015.pdf.

Love, M.S., M.M. Nishimoto, S. Clark and A.S. Bull, 'Renewable Energy in situ Power Cable Observation', US Department of the Interior, Bureau of Ocean Energy Management, Pacific OCS Region, Camarillo, CA. OCS Study 2016–008, at: http://www.boem.gov/2016-008/.

Malphrus, S., 'Undersea Cables and International Telecommunications Resiliency', 34th Annual Law of the Sea Conference, Center for Ocean Law and Policy, University of Virginia, 20 May 2010.

Manoj, C., A. Kuvshinov, S. Neetu and T. Harinarayana, 'Can undersea voltage measurements detect tsunamis?' (2006) 58 *Earth Planets Space* 1–11.

Marine Regions.org, 2016. *World maps* at: *http://www.marineregions.org/maps.php?album=3264&pic=64930.*

Marra, L.J., 'Shark bite on the SL submarine light wave cable system: History, causes and resolution' (1989) 14 *IEEE Journal Oceanic Engineering* 230–237.

Marshall, J., and K. Speer, 'Closure of the meridional overturning circulation through Southern Ocean upwelling' (2012) 5 *Nature Geoscience* 171–179.

McCave, I.N., and L. Carter, 'Sedimentation beneath the Deep Western Boundary Current off northern New Zealand' (1997) 44 *Deep-Sea Research* 1203–1237.

Monastersky, R., 'The next wave' (2012) 483 *Nature* 144–146.

MPAtlas, 2016. *How much of our ocean is protected?* (Marine Conservation Institute: Seattle, WA., 2016) at: http://www.mpatlas.org./.

Müller, C., R. Usbeck and F. Miesner, 'Temperatures in shallow marine sediments: Influence of thermal properties, seasonal forcing, and man-made heat sources' (2016) 108 *Applied Thermal Engineering* 20–29.

NASA, 2017. *Ocean Colour Web*; *Data* at: https://oceancolor.gsfc.nasa.gov/.

Natural History Museum, *Treasures of the Natural History Museum* (Natural History Museum: London, 2008).

NESDI, 'Studying the impact of seafloor cables on the marine environment' (Spring 2014) *Currents* 6–21, at: http://greenfleet.dodlive.mil/currents-magazine/currents-magazine-2014/currents-spring-2014/.

Ninety-Four Consortium Cable Owners v Eleven Named French Fishermen (http://www.iscpc.org/members/Legal/ Precedence_Cases/Boulogne_Judgement_Iss_1.pdf), Tribunal de Grande Instance de Boulogne Sur Mer (1st Chamber) 28 August 2009 (France).

NOAA, Commercial Fisheries Statistics—U.S. Foreign Trade, www.st.nmfs.noaa.gov/commercial-fisheries/foreign-trade.

Nordquist, M.H. (ed-in-chief), *United Nations Convention on the Law of the Sea 1982: A Commentary*, Vol. II (Martinus Nijhoff: Dordrecht/Boston/London, 1993).

Normandeau, E., T. Tricas and A. Gill, 'Effects of EMFs from Undersea Power Cables on Elasmobranchs and Other Marine Species', US Department of the Interior, Bureau of Ocean Energy Management, Regulation, and Enforcement, Pacific OCS Region, Camarillo, California, *OCS Study BOEMRE* 2011–09, at: http://tethys.pnnl.gov/publications/effects-emfs-undersea-power-cables-elasmobranchs-and-other-marine-species.

Ocean Biogeographic Information System, 2016. *Where is what and what is where?* at: http://news.coml.org/descrip/imgs/OBIS_marbef_articl03.05.pdf.

Ocean Networks Canada, 2016. *Discover the Ocean, Understand the Planet: Observations* at: http://www.oceannetworks.ca/.

Ocean Observatories Initiative, 2016, at: http://oceanobservatories.org/observatories/.

Ocean Observatories Initiative, 2017. *The Observatory*, at: http://www.interactive oceans.washington.edu/story/.

Organisation for Economic Co-Operation and Development, Broadband and the Economy, Ministerial Background Report, DSTI/ICCP/IE(2007)3/FINAL, 17–18 June 2008.

Palmer-Felgate, A., N. Irvine, S. Ratcliffe and S. Sui Bah, 'Marine Maintenance in the Zones—a Global Comparison of Repair Commencement Times', 2013, at: http://www.suboptic.org/document/marine-maintenance-in-the-zones-a-global-comparison-of-repair-commencement-times/.

Peracomo et al. v Sociéte Telus Communications, Hydro Québec, Bell Canada v Royal and Sun Alliance Insurance Company of Canada, 2012 FCA 199 (29 June 2012), *aff'd* 2011 FC 494 (2011) (Canada).

Pope, E., P.J. Talling and L. Carter, 'Which earthquakes trigger damaging submarine mass movements: insights from a global record of submarine cable breaks?', 2016 *Marine Geology*, at: http://dx.doi.org/10.1016/j.margeo.2016.01.009.

President Ronald Reagan Statement on United States Ocean Policy, 19 *Weekly Comp. Pres*, doc. 383 (10 March 1983).

Quintillion, 2016, *System- Phase 1 Alaska* at: http://qexpressnet.com/system/.

Renault, L., 'The Protection of Submarine Telegraphs and the Paris Conference (October–November 1882)', in *International Law Review*, ed. Brussels and Leipzig, (Flander: Merzbach and Falk, 1884).

Roach, J.A., and R.W. Smith, *Excessive Maritime Claims*, 3rd edn, (Martinus Nijhoff: Leiden/Boston, 2012).

Sandwell, D.T., R.D. Müller, W.H.F. Smith, E. Garcia and R. Francis, 'New global marine gravity model from CryoSat-2 and Jason-1 reveals buried tectonic structure' (2014) 346 *Science* 65–67.

Sherwood, J., S. Chidgey, P. Crockett, D. Gwyther, P. Ho, D. Strong, B. Whitely and A. Williams, 'Installation and operational effects of a HVDC submarine cable in a continental shelf setting: Bass Strait' (2016) 1 *Australia Journal of Ocean Engineering and Science* 337–353, at: http://www.sciencedirect.com/science/article/pii/S2468013316300316.

Smith, H.D., J.L. Suárez de Vivero and T.S. Agardy, *Routledge Handbook of Ocean Resources and Management* (Routledge: London/New York, 2015).

Starosielski, N., *The Undersea Network* (Duke University Press, 2015).

Stommel, H., 'The abyssal circulation' (1958) 5 *Deep-Sea Research* 80–82.

Submarine Cables and Deep Seabed Mining: Advancing Common Interests and Addressing UNCLOS "Due Regard" Obligations, ISA Technical Study: No. 14 (International Seabed Authority: Kingston, 2015).

Submarine Telecoms Forum, Inc., *Telecoms Industry Report 2012*.

Talling, P.J., C.K. Paull and D.J.W. Piper, 'How are subaqueous sediment density flows triggered, what is their internal structure and how does it evolve? Direct observations from monitoring of active flows' (2013) 125 *Earth-Science Reviews* 244–287.

TeleGeography, 2016. *Submarine cable map*, at: http://www.submarinecablemap.com/.

The Government of the Netherlands, Post Office v G'T Manneteje-Van Dam [Fishing Cutter GO 4], File No. 325/78 (District Court Rotterdam, decision rendered 20 November 1978), aff'd *sub nom* G.'t Mannetje-Post Office, File No. 69 R/81 and File No. rb 325/78 (The Court at the Hague, Second Chamber, decision rendered 15 April 1983) (Netherlands).

The Novorossiisk, *Dept. of State Bull.*, Vol. XL, no. 1034, at 555 (20 April 1959).

'The See-through Sea', *The Economist*, 16 July 2016.

UN doc. A/70/74 (30 March 2015), Oceans and the Law of the Sea Report of the Secretary-General.

UN, "Development of an Internationally Binding Instrument under the United Nations Convention of the Law of the Sea on the Conservation and Sustainable Use of Marine Biological Diversity of Areas Beyond National Jurisdiction," General Assembly Resolution A/RES/69/292 (2015) UN doc. A/RES/69/292: Development of an international legally-binding instrument under the United Nations Convention on the Law of the Sea on the conservation and sustainable use of marine biological diversity of areas beyond national jurisdiction.

UN, "Ethical Imperatives for a Nuclear-weapon-free World," General Assembly Resolution A/RES/70/50 (2015).

UN, "Oceans and the Law of the Sea," General Assembly Resolution A/RES/70/235 (2015) UN doc. A/RES/70/235 (23 December 2015).

UN, "Oceans and the Law of the Sea," General Assembly Resolution A/RES/69/245 (2014).

United Nations Convention on the Law of the Sea, 10 December 1982, 1833 *UNTS* 396.

United Nations World Ocean Assessment, 2017. *Report Outline, Completed Working Papers and Chapters* at: http://www.worldoceanassessment.org/?page_id=14.

US Geological Survey, 2016. *Earthquake Lists, Maps and Statistics*, at: http://earthquake.usgs.gov/earthquakes/.

Watkins, W.A., M.A. Daher, K.M. Fristrup, T.J. Howald and G.N. Di Sciara, 'Sperm whales tagged with transponders and tracked underwater by sonar' (1993) 9 *Marine Mammal Science* 55–67.

Whiteman, M.M., *Digest of International Law*, Vol. 9 (US State Department: Washington DC, 1968), at 948 (Alex Pleven) (France) [FN 84 format].

Wood, M.P., and L. Carter, 'Whale entanglements with submarine telecommunication cables' (2008) 33 *IEEE Journal of Oceanic Engineering* 445–460.

World Bank: "An Information-communication Revolution in the Pacific," World Bank Press Release (May 18, 2013), http://www.worldbank.org/en/news/feature/2013/05/17/information-communication-revolution-in-the-pacific.

Wright, I.C., 'In situ modification of modern submarine hyaloclastic/pyroclastic deposits by oceanic currents: an example from the Southern Kermadec arc (SW Pacific)' (2001) 172 *Marine Geology* 287–307.

WWF (Worldwide Fund for Nature), 'Taking a Risk Management Approach to Strategic Environmental Assessments and Environmental Impact Assessments for BBNJ', WWF submission to 2nd PrepCom, August 2016, at: http://www.un.org/depts/los/biodiversity/prepcom_files/WWF_BBNJ_Prep_Com2_2016.pdf.

Annex: Submarine Cable Repairs Worldwide 2008–2015

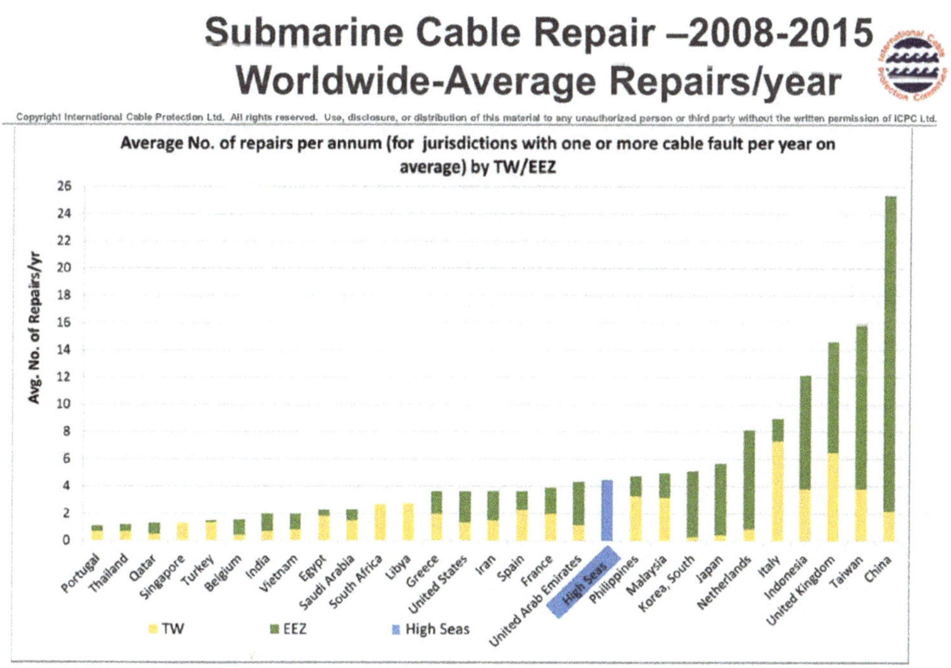

(Note that data for the United States includes repairs not only in the TW and EEZ of mainland USA but in its 'Unincorporated Territories' of the Territory of Guam and the Commonwealth of the Northern Mariana Islands (CNMI))

Submarine Cable Repair- 2008-2015 Average Time to Begin Repair

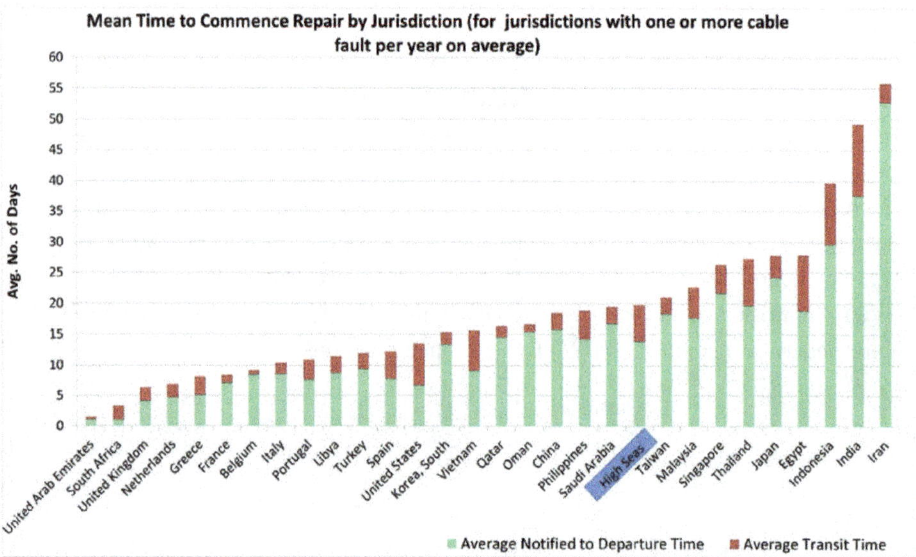

Submarine Cable Repairs-2008-2015 Causes for Repair Delay

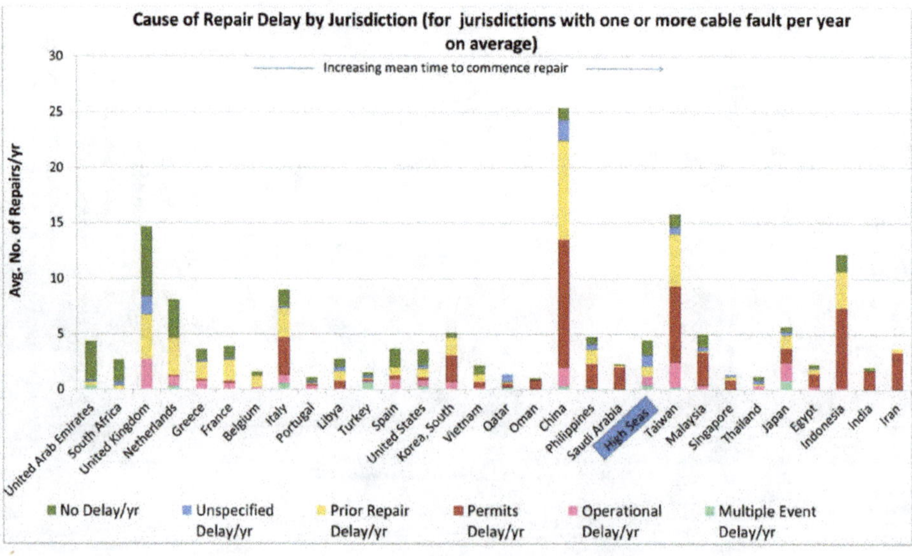

www.ingramcontent.com/pod-product-compliance
Lightning Source LLC
Chambersburg PA
CBHW071358230926
51846CB00002B/591
* 9 7 8 9 0 0 4 3 5 1 5 9 2 *